VEGETABLES LOVE FLOWERS

VEGETABLES
LOVE
FLOWERS

WITHDRAWN

COMPANION PLANTING FOR
BEAUTY AND BOUNTY

Lisa Mason Ziegler

COOL
SPRINGS
PRESS

Brimming with creative inspiration, how-to projects, and useful information to enrich your everyday life, Quarto Knows is a favorite destination for those pursuing their interests and passions. Visit our site and dig deeper with our books into your area of interest: Quarto Creates, Quarto Cooks, Quarto Homes, Quarto Lives, Quarto Drives, Quarto Explores, Quarto Gifts, or Quarto Kids.

© 2018 Quarto Publishing Group USA Inc. Text © 2018 Lisa Ziegler

Photography by Bob Schamerhorn, except where otherwise noted.

First published in 2018 by Cool Springs Press, an imprint of The Quarto Group, 401 Second Avenue North, Suite 310, Minneapolis, MN 55401 USA. T (612) 344-8100 F (612) 344-8692 www.QuartoKnows.com

Cool Springs Press titles are also available at discount for retail, wholesale, promotional, and bulk purchase. For details, contact the Special Sales Manager by email at specialsales@quarto.com or by mail at The Quarto Group, Attn: Special Sales Manager, 401 Second Avenue North, Suite 310, Minneapolis, MN 55401 USA.

10 9 8 7 6 5 4 3

ISBN: 978-0-7603-5758-3

Library of Congress Cataloging-in-Publication Data

Names: Ziegler, Lisa Mason, author.
Title: Vegetables love flowers : companion planting for beauty and bounty / Lisa Mason Ziegler.
Description: Minneapolis, MN : Cool Springs Press, 2018. | Includes index.
Identifiers: LCCN 2017043293 | ISBN 9780760357583 (sc)
Subjects: LCSH: Companion planting. | Vegetables. | Flowers.
Classification: LCC SB453.6 .Z54 2018 | DDC 635—dc23
LC record available at https://lccn.loc.gov/2017043293

Acquiring Editors: Mark Johanson and Madeleine Vasaly
Project Manager: Alyssa Lochner
Art Director: Cindy Samargia Laun
Cover and Interior Design: Amy Sly

Printed in China

To my little sis and best friend, **Suzanne**.
I love that we do this together.

TABLE OF CONTENTS

Introduction

It All Comes Down to Flowers

A summer harvest. Clockwise from top left: celosia 'Chief'; black-eyed Susan 'Triloba'; zinnia 'Benary's Giant'; tomatoes 'Black Cherry', 'Cherokee Purple', and 'Big Beef'; and Mexican sunflowers.

What business do pretty flowers have in a vegetable garden? Not much, many gardeners might think—in fact, I married into a large vegetable-gardening family, and that's what they thought too. Flowers were a waste of precious space and labor in the vegetable patch. Treated like tagalong little sisters, they got a place only if there was room leftover. But all that changed with the discoveries I made on the way to becoming a cut-flower farmer.

Flowers have been a key element of the vegetable patch for centuries. The cottage garden is one example of the practical use of small-space gardening to grow food crops, with a healthy dose of flowers on the side to attract pollinators and beneficial insects. Today, flowers are often a casualty of downsizing and practicality, but in fact, flowers more than pull their weight in the garden! Ornamental and functional at the same time, they put out the welcome mat for pollinators and beneficial insects. The guests will come, set up house, and raise their families—exactly what you need in a healthy garden.

After years of gardening and farming, visiting countless gardens planted by others, and fielding

SPRING: *Snapdragons 'Rocket' and 'Madame Butterfly', sweet William 'Amazon', dill 'Bouquet', bupleurum, and false Queen Anne's lace 'Dara'.*

SUMMER: *Sunflower 'ProCut', zinnia 'Benary's Giant', celosia 'Sylphid', celosia 'Jura Salmon', cosmos 'Double Click', tansy, black-eyed Susan 'Triloba', and false Queen Anne's lace 'Graceland'.*

thousands of questions, there is one thing I have seen time and time again: allowing nature to provide the pest control, pollination, and nutritional systems in our gardens is so basic that we don't even think about it. It just seems too simple to work. But work it does when we give it some flowers and a chance.

Everything began to change in my garden when I added three seasons of pesticide-free flowers. Through the years, as I coupled the presence of flowers in the vegetable garden with common-sense natural gardening practices, my garden flourished. What rose above any gardening troubles along the way was that my garden filled up with pollinators and nature's pest controls—especially welcome at a time when the number of beneficial insects in most gardens, especially bees, has been diminished by pesticide exposure and loss of habitat. Gardening became easier and my harvests more abundant.

The practice of keeping a small cut-flower garden within the vegetable garden is nothing short of delightful. The gardener gets an armload of beautiful fresh-cut flowers each week, and the beneficial creatures are attracted to and happy with the continuous supply of new blossoms. Those beneficial creatures come for the flowers, then share their benefits with the nearby vegetables. The routine of harvesting the flowers for the table keeps the cutting garden alive and producing. This in turn keeps the garden full of fresh flowers for all. Just as we

FALL: *Sunflower 'Moulin Rouge' and 'ProCut', Mexican sunflowers, celosia 'Jura Salmon' and 'Sunday', and 'Cinnamon' basil.*

harvest the vegetable garden to keep it producing, we do the same with flowers.

I didn't set out to fill my gardens with pollinators, beneficial insects, and other good things. You might say they all came as a welcome side effect of all-natural cut-flower farming. I finally gave up messing with the ecosystem and instead gave it a hand up and helped it along. The result was a garden teeming with beautiful healthy plants producing abundance with little intervention from me, a garden just the way it was meant to be.

The view from my office desk overlooking the garden. The plants thrive because their roots run deep in our well-cared-for soil.

We grow vegetables on our flower farm for our farm families and cut-flower customers. These are early-summer beds of bush beans and tomatoes. PHOTO BY GARDENER'S WORKSHOP FARM

MARRYING INTO VEGETABLE GARDENING

Soon after marrying into this gardening family, I took over the vegetable growing on the family homestead that was now my home with my husband, Steve. There were two large vegetable-garden areas, each a quarter acre. The tradition was to grow corn, beans, peas, tomatoes, potatoes, and squash—and plenty of them—for eating fresh and for storing up for out-of-season use for our larger extended family.

I was woefully inexperienced. I had done a little shade gardening in a small space, but I knew nothing about vegetable gardening. I had never grown anything that required full sun. While I was over the moon with excitement to take on this task, I really had no idea how to go about it.

With Steve's help and encouragement, I forged ahead. I began to learn the basics of row vegetable gardening, the traditional style of the day. I learned when to plant and how. I learned how to use the walk-behind tiller to prepare and maintain the garden. I had so many questions: Where do you purchase seed when you need

more than a packet? What do you do about weeds? What about watering—how and how frequently? The learning curve went on and on and on. It was a challenge, but fortunately for me, I felt invigorated, not overwhelmed.

As might be expected for a novice, that first vegetable-gardening season was a doozy. I had some successes, then there were the disasters. The work seemed endless, but the full harvest baskets outweighed all the sweat equity. Looking back, I realize I had the benefit of the blissful blindness of inexperience as well as the surprising ease with which everything seemed to grow. They carried me through those first gardening years.

From the start, I was growing vegetables in such abundance that it was easy to overlook pest and disease red flags—until I faced my first major pest outbreak. I was devastated. String beans were among the staple vegetables I grew then and still are today. By the time I noticed the damage, the leaves of the entire crop were riddled with holes. My plants hadn't even begun to bloom yet, much less

I am able to grow string beans with little to no damage from the Mexican bean beetle now that I follow some specific growing habits to help deter them.

produce any beans. Upon closer inspection, I could see burnt-orange beetles with black spots crawling everywhere: Mexican bean beetles.

It pains me to remember and write that I treated those beans with a pesticide, but Steve's gentle wisdom opened my eyes to another way. What he and his family had learned over generations of gardening was to beat those beetles with timing. Planting as early as possible in the season would outsmart the pests. In my busyness, I had disregarded said advice and had planted late. Planting bean seed as soon as the soil is warm gives the plants a chance to mature and begin bearing beans before the beetle outbreak, thus avoiding much damage. Looking back, I can see that Steve's advice was my first lesson on how to use nature to benefit the garden.

This lesson of simple timing laid the groundwork in my mind of the potential benefits to the garden and gardener of an assortment of garden practices. I began to question when to plant a given plant, when to mulch, or when to run a hoe throughout the garden. I suspected that many of these activities offered benefits based simply on when you did them. You aren't doing additional chores—you are just aiming to do them when there is maximum benefit. Getting the timing right for each chore is key to a healthier garden and lower maintenance.

Another part of my story that started to come into focus at this time had to do with my garden soil. There was something very special about it. Everything I planted grew really well in spite of my lack of know-how. I

thought it must be its history of being treated as a garden for so long before I arrived.

And then I found the stacks.

One of the really exciting things about living on a family homestead is discovering history. There were several outbuildings when I came to live here. Two of them are still fully functional, but the other two were too badly damaged from time and weather to be saved. Steve and I set out to empty the buildings of decades of family farming history and then moved on to disassembling them to save the wonderful boards. What a walk through history this was!

In one of those buildings slated for teardown, I came upon ceiling-high stacks of debris I didn't recognize. It appeared that a raccoon had been living in the piles for quite some time. Steve joined me and immediately recognized what they were. His grandpa had collected thousands of bags of leaves from neighborhood curbsides over the years. He made leaf mold with the leaves and then added it to the garden soil. Those towering stacks were the emptied leaf bags he kept to reuse, as anyone that grew up during the Depression era would have done.

Bush string-bean foliage riddled with holes made by the Mexican beetle. Once the damage is this widespread there is little that can be done.

That's the day the tide began to turn on my gardening point of view. I cannot guess how many empty bags lay there as evidence of how much leaf mold had been made and turned into the garden soil I now worked. I was perplexed over these findings—the soil eats leaf

One of the surviving outbuildings on the farm. The "Inn" served as a cottage back in the day and is now my sweet potting shed.

mold? Does that mean it is alive? Was this the reason the garden produced abundance so easily? And what had it had to eat since grandpa's gardening days?

Up until this point, I had been planting and tending and occasionally treating for pests with pesticides. I had total disregard for the soil I stood on. All of my attention had been focused on the aboveground part of the plants. Little did I know that it's what the roots live in that controls everything! That it is the foundation of a plant and key to the plant's existence and state of health. What I know now is that the performance of the plant aboveground is a direct reflection of what is going on below-ground. Having garden troubles and struggles? Chances are good you are standing on the problem.

With so many questions, I plunged ahead to learn more, and that included enrolling in the master gardener program. I credit that course with sparking my desire to learn more about the soil. I discovered there that soil is in fact alive and full of billions of microorganisms

and that they eat leaf mold, compost, and other organic matter. They spend their lives nurturing, feeding, and caring for plant roots. While we don't need lots of scientific knowledge to keep a healthy garden, we do need to know that feeding the microorganisms should be a top priority. When these guys go hungry and cease to exist, our garden will be full of dead dirt.

In one of those training classes, a speaker from the Chesapeake Bay Foundation shared the horrendous polluted state of the bay and how it had come to be that way. It became crystal clear to me that what we do in our gardens doesn't stay in our gardens. The gardener's way of gardening has the potential to affect so much beyond the garden gate, both for good and bad. I knew then that I would do my part not only to *not* cause harm but also to help restore the natural order of my garden and do it while producing abundance. At this point, I still had no idea how much impact flowers had in the natural order—I just knew I wanted to be a flower farmer.

ADDING FLOWERS

The best dose of medicine my gardens ever received were flowers. Looking back, it was a perfect storm of circumstances that unfolded and ushered me into flower farming: my beginner's success growing vegetables, the light-bulb moment on how nature could help my garden, and those stacks of empty bags. And all the while, I was putting my dream into reality—growing flowers as my chosen work.

I launched into growing cut flowers to sell in 1998 and was met with the same growing success as I had had with vegetables. Everything I planted grew like mad. At first I added a bed or two of flowers within our large vegetable gardens, and then as the demand and my business grew, I became a full-time urban flower farmer. During this time, my gardens grew from two ¼-acre gardens to also include an additional 1-acre garden.

It didn't take long for my gardens to start filling up with some obvious good things beyond the flowers I was now planting. As soon as the blooms began, it seemed that butterflies, bees, and birds were everywhere. There among the flowers were creatures buzzing and flitting around just like you imagine nature at its best.

I grew zinnias, sunflowers, snapdragons, sweet peas, and other popular cut flowers. They were much more than just pretty faces—they provided food and habitat for beneficials. While there are flowers that are especially attractive to specific groups of beneficial insects, the bottom line is just is to grow flowers—any flowers!

Elephant garlic grows next door to poppies. The pollinators appreciate all the blooms, evident as a honeybee comes in for a landing.

An adult soldier beetle on a giant marigold. In their larval stage, soldier beetles are important predators of many insect pests, including grasshopper eggs.

Don't use pesticides and they will come. That is exactly what I did.

What brought the garden to life and kept it blooming over the long haul of the seasons was growing flowers for cutting and then harvesting the blossoms on a regular basis. Constant cutting keeps the garden continually producing fresh flowers. Fresh flowers keep all the beneficial creatures happy in the garden. As it turns out, the life of a cutting garden is a perfect match for the life you wish to invite into the garden.

This practice of harvesting flowers while harvesting vegetables brings another gift from the garden. The day will come when there isn't room for another bouquet in your home. This leads to a new reward—surprising and delighting family and friends with the colorful gift of fresh, homegrown flowers!

An early-spring harvest is easy when the vegetables, herbs, and flowers all share the same space. Here, pot marigolds, kale, parsley, squash, and lettuce.

The stakes throughout my garden make a perfect perch for insect-eating birds. Here is a mockingbird wondering what we are doing in his garden.

THE RESULTS OF FLOWERS

As my garden became home to more and more flowers and beneficial creatures, I became concerned about using any pesticides, organic or otherwise, that might have effects beyond the targeted problem. I started reading labels from a different perspective. Instead of being eager to know what problem a product could solve, I was suddenly wondering what it might harm beyond the problem, including the good things now living in my garden.

While I was doing my best to find a balanced way to safely treat for pests and diseases, my little flower-farming operation really took off. What happened next was not an educated guess or even a plan, I confess. It just happened because I got so busy and overwhelmed with farming flowers and vegetables that I simply had no time or energy to do anything about pests. I quit all treatments. I abandoned the garden to let it solve its own problems. That turned out to be one of the best things that ever happened to me and my garden.

An outbreak of aphids on a milkweed plant that will be followed by a stampede of ladybugs and other pest-eating good bugs. Pictured is Gomphocarpus physocarpus.

Pest outbreaks were happening all along, but it appeared that the garden was resolving each problem or at least controlling it to an acceptable level. It was dawning on me that a low level of pests in the garden is actually a good thing. After all, for the beneficial insects and creatures to stay in the garden, they must have a source of food. Those pests were the actual fuel for the fires of nature to work.

I began to notice fascinating things happening in my garden. I understand now why they make animated movies about insects—my earliest memory of witnessing predator insects taking out a pest is forever ingrained in my mind. I was on the ground in a pathway between two beds, weeding. As I was working, some movement caught my eye up on the stem just below a zinnia bloom. Just as I turned to watch, a little posse of orange-bodied, long-black-legged bugs were running down the stem in single file. They snatched up a caterpillar, struggled with it for a moment, and then carried it back up the stem underneath the bloom. They proceeded to consume it. *Whoa*! I was practically out of breath when this was over.

Nature is very impressive. These garden warriors were assassin bugs. Once I read up on what they were and what they do in a garden, I knew I wanted more. Those I saw that day were the nymphs, or teenagers, that weren't able to fly yet. The assassin-bug family is large, with over 160 species in North America. They are predators that aren't picky about what they eat. They will go for small insects such as aphids as well as those giant caterpillars chowing down on your tomato plants.

An assassin bug going after a moth caterpillar. Assasin bugs will attack both pests and beneficial insects in the garden and, if provoked, will bite humans.

Intrigued? I was. The assassin bug is just the tip of the iceberg of nature's workers that will work and live alongside you to keep your garden in a natural balanced ecosystem. And there is more good news, according to the University of Florida Extension: of the 100,000 insects found in North America, less than 1 percent of them feed on plants in a harmful way. Let that sink in for a moment. That means that the vast majority of bugs in your garden are most likely not harmful and have a good chance of even benefiting the garden in some way.

There are some beneficial creatures that you are likely acquainted with but probably don't welcome. The primary three I'm speaking of are snakes, wasps, and spiders. I had to get over my fear of these because they are such important players in pest control in the garden. There was no one more afraid of a wasp than me, but believe it or not, you may come to love them as much as I have. When you learn that the harmless snake hanging out in the rock pile just swallowed the vole (garden mouse) that ate all your tulip bulbs, or when you find

stinkbugs wrapped up in a spider's web, the rewards of revenge can be sweet.

After working so hard to bring the garden to life, to choose to let a problem go is not the first natural reaction. Finding pests or damage in your garden throws you for a loop, and that little "oh no" slips out. Even now, when I encounter a problem in my garden of any kind, it's like a personal insult—a punch in the gut. My instinct is to intervene immediately, but then I come back around to my newfound gardening senses—I push myself to walk away and give nature time.

This discovery of what nature has to offer still continues for me today, and I enjoy learning more about this little-known community of beneficial creatures and what they do to help my garden thrive naturally. Some are so small you can barely see them; many are below ground level working invisibly, and others may have even terrified you in the past. This garden community is as diverse as the population of any major city. Each creature has a job and fills a need in the chain of life in a garden.

A wasp and ladybug helping to control the pest aphids on the milkweed. Wasps are one of the most effective pest eaters on my farm.

Summer brings abundant harvests of sunflowers, tomatoes, beans, cucumbers, and figs. What we can't eat or preserve, our flower customers purchase.

A PATHWAY TO EASIER GARDENING

Because I planted flowers on such a grand scale, it was impossible not to notice all the good things the flowers were bringing to the table. It felt as though nature hit me right between the eyes with all it had to offer. I had expected a harvest of beautiful blossoms but never dreamed they were also the key to healing my garden, bringing in entertainment, and making gardening easier! The garden grew with less help from me with each passing season.

Chatting with gardeners through the years, I've learned that the benefits of flowers in the vegetable garden often haven't been given a chance to develop. Perhaps a large vegetable patch only had a small corner for flowers. Perhaps deadly pesticides were being used. Perhaps flowers were planted in such a way that blossoms were present only sporadically throughout the growing season. These factors may cause the benefits to either vanish or just limp along. Adding a balance of blooming flowers to vegetables gives nature a chance to develop into something, and that something is spectacular!

I continue to grow both vegetables and flowers, hand in hand, sharing the same spaces. As I witness nature at work in my garden, it has become easier to walk away from pest problems. All the energy I had been using to resolve problems is now focused on learning more about how to help the garden in order to prevent problems.

Adding flowers to the vegetable patch and following nature's lead is not a magic bullet. It will not fix your garden problems overnight, but it will heal a garden over time. While there are many good reasons to go natural in the garden, the resounding reason that every gardener can get on board with is that it just makes gardening easier and more abundant.

In these pages, I will unravel how easy and fun it can be to use nature to prevent and combat problems. My garden today is planned around nature's workforce as much as it is for the vegetables for our families and our cash crop of cut flowers.

When flowers are coupled with all-natural gardening practices, it changes everything. It's not just adding flowers or stopping the use of pesticides or any other one thing—it's all those things working together that will invite nature in and let it do what comes naturally to help your garden grow.

Our tomato plantings are spread out amongst the flower plantings to help deter disease and pest issues—including deer. Tomato blossoms are as beautiful as our cut flowers.

How Gardening Gets Easier

- **The Vegetables:** Planting vegetables into their favored conditions and at the right time leads to stronger, more productive plants with fewer pest issues.

- **The Flowers:** Including a strong dose of flowers from early spring to frost will attract and keep nature's workers year-round in the garden.

- **The Soil:** Adding organic matter to the garden soil on a regular basis will build and sustain healthier garden soil with each passing season.

- **The Plants:** Starting with healthy plants has an important impact on the life of the garden. This means starting from seed whenever possible.

- **Day-to-Day Living:** Prevention is key. Spending a little time setting up the garden to prevent issues such as weeds will save a whole season of chores.

- **Providing Habitat:** Giving the beneficial creatures a place to live, eat, and raise families means they will stay in or close to the garden, along with their offspring. Habitat is all about building community.

- **The Movers and Shakers:** Learning to make and have compost and leaf mold on hand as part of the garden's landscape makes caring for the soil easier.

The Battle on Mexican Bean Beetles

I've won the war. The season after my string beans were attacked, I began doing everything I could to stack the cards in favor of the beans, not the beetles. I studied the life cycle of the beetle, and it was like reading what had unfolded in my garden that first season. It may have seemed to happen overnight, but it did not.

While the unprotected plants were growing, the beetles flew in from their nearby habitat. They fed on the bean-plant leaves for 1 to 2 weeks and then began laying eggs, up to 600 eggs per adult. Those initial adults caused some damage, but the devastating damage came when all those hungry babies hatched and started eating. By midsummer, the second generation was well on its way and caused the greatest damage.

After that, I followed Steve's advice and planted the bean seed as soon as the soil was warm enough. I also immediately covered the seedbed with a floating row cover until the beans began to bloom. The cover allowed air, water, and light to penetrate but kept the beetles out and delayed their life cycle. Keeping my eyes peeled for beetles and handpicking any beetles or larvae went a long way. Each one eliminated potentially prevented hundreds of beetles down the road. Warning: handpicking can become obsessive when you consider what an impact it can have on pest control!

My bean crops now have practically zero beetle damage because of a combination of things: the presence of their natural predators invited in by flowers, planting timing, the delay of exposing plants to egg-laying adults, and handpicking. Nature and practices worked together to prevent and control the problem.

TOP LEFT: *A bed of healthy midsummer beans free from Mexican bean beetles and their damage—something I could have never mastered before I started working with nature.* TOP RIGHT: *One of nature's warriors in action. The beneficial spined soldier bug attacking Mexican bean-beetle larvae.* PHOTO BY USDA BOTTOM LEFT: *The adult Mexican bean beetle that causes devastation to snap bean plants. They not only feed on the plants but can also spread viruses.* PHOTO BY STEPHEN AUSMUS, USDA BOTTOM RIGHT: *Young bean plants that were covered with a row cover immediately after the seeds were planted. The cover is kept on at all times to prevent access to Mexican bean beetles.*

SECTION 1

Flowers in the Vegetable Garden

When you bring flowers into your vegetable patch, be prepared for good things to happen. With a small cutting garden tucked in with the vegetables, it is easy to keep flowers blooming for the beneficial insects, pollinators, and other good things. The fresh bouquets are a bonus for you! Harvesting the flowers regularly just as you do vegetables keeps the plants producing blooms throughout the growing season. A little-known fact about cutting gardens is that they don't need much space because the continuous harvesting stimulates so many fresh blossoms. It still surprises me how many flowers a small area can produce.

As you become familiar with all the benefits flowers have to offer, you will be eager to get them started in your patch. My suggested garden plans in the appendix are a jumping-off point to tweak to your needs. The plans demonstrate how to succession plant through the four seasons and to never leave the soil bare and exposed. The silver lining of this first section of the book is taking to heart how to treat a cutting garden. No matter where you garden, one truth holds true: the flowers in a cutting garden must be harvested on a regular basis to survive and thrive. The more you cut, the more they come, I promise!

OPPOSITE: *Squash, cucumbers, and tomatoes growing in harmony with flowers. Not only do the flowers attract bees and beneficial insects, they make the garden beautiful.*

Why Do Vegetables Love Flowers?

Beneficial insects are attracted to the garden by the flowers. With flowers grown alongside the vegetables, they don't have far to go to find pests.

I see and hear so much during all the hours I spend harvesting flowers on the farm—it hardly seems like work at times.

While you're in my garden you would never guess that just outside my property line is a bustling neighborhood located in the midst of a city.

I have been in a unique position the past 19 years as an urban cut-flower farmer. I spend 6 months of the year standing out in my garden for countless hours each week harvesting flowers and watching the day go by. I have witnessed things I would never have guessed were possible, much less happening in my own garden. As a farmer, gardener, birdwatcher, and general wildlife lover, I am surprised by what I see. And to think it all happens in the middle of a city of 200,000 residents.

All this action in my garden wasn't always the case. It came as I moved my focus away from "just" growing plants to focusing on beefing up the ecosystem so it could grow healthy, robust plants naturally. That's when the garden came alive. What does that mean? You could say that organic gardening is rooted in preventive medicine. I began to garden in advance of possible problems. I learned to do the chores that prevented issues instead of running after the problems, picking up the pieces and trying to fix them. It turns out to be an easier, less stressful, and more cost-effective way to garden. The truth is, I can't imagine having farmed all these years without following nature's lead.

My gardening experiences have unfolded in the middle of a neighborhood with school buses rattling by, neighbors' blowers going, and even the occasional city mosquito truck spraying its mist over the area. Nature can overcome much because it never stops trying. Pest insects seem to come so quickly when you plant their delight (like squash bugs to squash), but it is just as true for the good things of the garden—plant their delights and they too will come.

Without flowers, there is no reason for any of these good things to come, much less stay and call the garden home. The experiences I share are not exceptions or even every-now-and-then happenings; this kind of stuff is happening day in and day out right here in my city garden. In fact, I get so caught up in watching it unfold that at times my farm coworkers inquire whether we are working or watching nature today.

Because of the ladybug, gardeners everywhere have gained an understanding of how powerful beneficial insects can be and the impact they have in a garden.

COMPANY'S COMING

Adding flowers will likely bring you face to face with one of the most well-known beneficial insects, the ladybug. Show a photo to even nongardeners, and they are almost sure to recognize its cute little red body with polka dots. It is said that the ladybug came to fame because farmers believed it was sent to deliver them from the ravages of pest attacks back in the day. Regardless of how it came to be famous, the fact remains that the ladybug is a fantastic example of nature helping us solve a problem—one little ladybug can eat thousands of aphids and other garden pests in its lifetime. How's that for pest control?

The great news is that there's an army of other insects just like ladybugs that patrol the garden and solve problems. Some prey on pests and eat them, while others will parasitize the pest, eventually killing it. There are minute pirate bugs that, in spite of their tiny $\frac{1}{16}$-inch size, will consume thrips and mites. There are damsel bugs that will gobble up soft-bodied pests. The garden becomes home to a diverse mix of problem-solving beneficial insects that come for the flowers and stay for the pests. Without flowers, pests can run rampant and create a host of problems that lead our gardens astray.

Because I hang out in the garden so much, I can't help but notice the little dramas playing out on my plants and in my soil. Did you know that some wasps are meat eaters? They patrol the garden looking for caterpillars and other insects to sting and carry back to their nest for food. Those large black beetles that are crawling around the garden? They capture and eat soil-dwelling insects such as slugs, maggots, and wireworms. I hope it is becoming clear just how the garden can take care of its own business when given a chance.

If the promise of good bugs eating bad bugs isn't enough reason to give flowers some room in the vegetable patch, I have more. There's a community of ready and able pollinators that most folks hardly know exist.

Native bees, just like their honeybee relatives, are suffering at the hand of habitat loss and pesticide exposure. There are thousands of native bee species right outside our back doors. These bees play a major role as pollinators in the home garden. They also play a critical role in maintaining the beauty of our iconic landscapes by pollinating the plant world beyond our gardens.

Growing tomatoes? You will definitely want to invite the most well-known native bee of all to your garden, the bumblebee. They are excellent pollinators. Good pollination leads to larger, more abundant, and better-quality fruit. My garden is loaded with bumblebees, and I give them the credit for the spectacular bumper crops of tomatoes! They are easily attracted to the vegetable garden by way of flower blossoms.

BUMBLEBEES AND TOMATOES

What do bumblebees offer tomatoes that other bees don't? Buzz pollination. Tomato blooms have all the necessary parts to self-pollinate, so to get that process moving, either wind needs to knock the pollen loose or a bumblebee can visit and do the job. The bumblebee reaches in and grabs the anther with its jaws and vibrates its wings. This action forcibly expels pollen out, where it would have been otherwise trapped.

Bumblebees are essential to strong tomato pollination. Snapdragons, a favorite of bumblebees, attract them to the garden just in time to visit the tomato blooms.

The garden stakes provide a perch for this thrasher to hunt for insects and watch her babies. Do you see the whiskers below her beak?

BEING A GOOD HOST

As a home gardener, you might take some tips from the flower farmer on what supplies to use. Some have proven to benefit not only our plants but also the beneficial workers in the garden. This has led me to add other items over the years, such as habitat features. Here are some of our favorites.

Placing water in the garden will train birds and other creatures to drink from the vessel instead of pecking tomatoes and other fruits.

GARDEN STAKES

Hundreds of stakes stick up all over my garden. They hold up the netting that supports our flowers and pepper plants when they're loaded with fruit and blossoms. But that's not all they do. At times, there may be a bird perched on almost every stake—a welcome sight. Beyond their insatiable appetite for insect pests, birds are endlessly entertaining as they jockey for position.

Who else do I find hanging around on those stakes? Dragonflies, juvenile great horned owls, and red-tailed hawks. On occasion a brave squirrel might venture out into the garden to perch on a stake, but he doesn't try that without looking over his shoulder with our strong predator presence. Thanks to the predators, we no longer have problems with rabbits, voles, moles, or any other rodent pest in our garden.

TRELLISING

We use trellis netting to support crops such as runner beans and sweet peas. Its visitors vary with the seasons. It can become a king-of-the-mountain standoff at times.

A line of hummingbirds shows up in the early morning until the mockingbird arrives and starts talking their ears off. As the day heats up, doves and other birds pick spots along the trellis as they hunt for insects. It turns out they all enjoy a bird's-eye view of the garden and everything it has to offer.

IRRIGATION

Water is an essential element in the garden. We use lay-on-the-ground irrigation that is easy, reusable, and inexpensive. In the early days, I had far more operator-error leaks than I do today. How did I know I had a leak? Flocking birds! Another lesson I share in Chapter 8 is that much of the damage done by creatures in a tomato crop happens when they're in search of moisture. That's why they peck or bite a hole, suck out moisture, and leave the rest. Providing water to your garden visitors can reduce such damage.

BIRDHOUSES

Including houses around our garden has upped the ante on the volume of insects consumed by birds. Our eastern bluebirds have been raising broods each year for some time. It's the highlight of our season to watch them constantly swooping into the garden, snapping up insects, and carrying them to their mates or fledglings. How convenient that just as the insect pests start going full throttle, the hungry babies start hatching!

KEEPING BENEFICIALS WITH FLOWERS

Having continuous blooms in the garden throughout the growing season is important, but spring and fall flowers are especially crucial for attracting beneficials. If nothing is blooming in gardens and landscapes then, the beneficials are left high and dry.

IN SPRING

When the cool-season vegetables are going in the garden, plant cool-season flowers too. Next to fast-growing early radish seeds, plug in snapdragon transplants that will happily fill in the harvested radishes' space in the coming

Bachelor buttons are one of the first flowers to bloom in early spring. These tiny beneficial hoverflies flock to them because little else is available.

coming weeks. As special bonus, those bumblebees you want for your tomatoes adore snapdragons.

Flower blooms in the vegetable patch in early spring will jump-start the natural pest-control community. Due to habitat loss and the use of herbicides, the early-blooming wildflowers they have depended upon in the past are few and far between. The blooms of cool-season annuals such as bachelor buttons and calendula will provide food for the early arriving pollinators and beneficial insects.

I've learned to sow my early-spring peas alongside my bachelor-button flower patch. Before my flower-growing days, the peas always suffered from an outbreak of aphids just as they were ready to harvest. Now with the strong presence of beneficials, our peas rarely show any signs of aphids. Because beneficials are living and reproducing in the garden, there is a continuous supply of hungry babies that can prevent a massive outbreak.

Getting the beneficials into the garden earlier also gives their life cycles maximum benefit. Pest insects typically reproduce at a greater rate than beneficial insects. By getting the good guys established in the garden early, they start reproducing sooner. Once reproduction starts in the garden, the population of beneficials will grow and multiply.

Of course, there are also the beautiful spring bouquets. Because the underlying heartbeat of this garden is the constant harvesting of stems, you must harvest to keep the blooms coming at all stages. The charm of homegrown spring flowers on the table will compel even diehard vegetables-only gardeners to plan for and plant summer flowers.

IN SUMMER

It's the faithful act of harvesting the flowers and vegetables that keeps the bounty coming nonstop. Because summer harvesting alone is enough to make a gardener run away to never return, the summer garden is set up to produce with as few chores as possible. Planning ahead is how we survive brutal summers in the Southeast.

The heat and bright colors of summer flowers bring in another layer of creatures to the garden. Zinnias, sunflowers, and basil are some of the main attractions. Butterflies, bees, and hummingbirds start showing up in larger numbers. Hummingbirds will work the zinnia stems, even the ones we have harvested and hold in our hands. Harvest buckets often include a butterfly working a flower until we carry the bucket indoors. Many of the heat-loving summer flowers will continue to bloom into fall.

IN FALL

There are so many great fall bloomers that benefit the garden and the gardener. Cosmos love fall conditions, and a late-summer planting will have them putting out

Zinnia 'Benary's Giant' is available in mixed colors or in solid colors. Zinnias attract butterflies, bees, hummingbirds, and other beneficial creatures.

The fall-blooming salvias are a bumblebee favorite and a great addition to the fall garden. Pictured on the left is Salvia leucantha *and on the right is* Salvia mexicana. PHOTO BY GARDENER'S WORKSHOP FARM

blossoms for your table and for the pollinators. Another favorite is Mexican bush sage; bumblebees will come from far and wide to visit its velvet purple blooms. What may be most surprising for newbie cut-flower gardeners is the ability of summer flowers to continue blooming into fall. This happens because of faithful harvesting habits.

I plan ahead so that I always have something in bloom. But as the days get shorter in fall, all good things start to head into hibernation, fly south, or hunker down in the garden for winter. It is then that the habitat you have allowed to happen or provided will come into play to make next year's garden healthier and easier than this year's.

Tomatoes and squash grow happily alongside zinnia 'Benary's Giant Pink' (front) and zinnia 'Cupcake' (back).

THE BIG PICTURE

The truth is that setting up the garden to help take care of itself is rooted in flowers, but it takes more than just that. It is more than a single step that brings this all to life; it is a combination of steps that encourage and help nature along. Many gardeners have tried to grow organically but have faced mixed results with frustration and often will fall back on old ways. Let me be clear on this: using pesticides, organic or otherwise, can harm beneficial insects and undermine the very workers that flowers have attracted to the garden.

While I go deeper in Chapter 8 on alternative ways to handle and avoid pest outbreaks, here are some reminders. To get your garden on its way to solving its own pest issues, you must give it time to work. Season after season of pesticide-free gardening will result in a large community of naturally occurring beneficial insects. Natural pest controls are more of a marathon than a sprint. If you find aphids in the garden, give the beneficials a chance to find them. They will come, eat, mate, and lay eggs, but you must give them time.

After following natural gardening practices, you will find the beneficial community becoming more mature and widespread. Major pest outbreaks will become a thing of the past as the beneficials are always present and find the pests before they get out of hand. Problems will often be resolved without your knowing that there was ever a problem.

There is so much good that starts to happen in the garden when you go all natural and add flowers. It helps the garden spark even more joy, and it spreads the joy by way of the beautiful cut flowers.

It is the "cutting" in the cutting garden that keeps the flowers producing. The flowers then attract the good things we want and need in the garden. Behind the scenes of those flowers, a world unfolds that few are aware of. It is fascinating, beneficial, and useful, just the way a garden was meant to be.

As a commercial farmer, I make regular harvesting a way of life. The secret to abundance in the flower garden, the vegetable garden, and the herb garden is the same—harvest often and you will be rewarded. Read on for more good reasons to restore the natural order of your garden and bring nature in to do the heavy lifting on insect control and more. It doesn't stop there.

A spring harvest of snapdragon 'Chantilly' and 'Madame Butterfly', poppy 'Champagne Bubbles' and 'Giant Pods', bachelor buttons 'Boy', bupleurum, pot marigolds 'Pacific Beauty', dill 'Bouquet', and false Queen Anne's lace.

How to Interplant Your Vegetable Patch

Bringing your vegetable garden to life with flowers is easy. Designating two small spaces in a vegetable patch for flowers and then treating them as cutting gardens will produce a steady supply of flowers from early spring right up to the first fall frost. This little flower-garden island will provide more flowers than you might expect for both you and the beneficial creatures you want to invite into the garden.

The most productive vegetable gardens are made up of annual plants, and the same is true for cut-flower gardens. Annual plants go from seed to producing blooms and fruit within 1 year. These plants produce the most per square foot.

A bonus of growing annual plants in the garden is the clean-slate effect: you can try new and different plants each season. While there is a place in a cutting garden for perennial plants, such as the peony, I find perennials best suited to the surrounding landscape or other designated garden spaces, not in the annual bed or garden.

The basic recipe I follow in the garden is 40 percent flowers to 60 percent vegetables. This ratio allows enough space even in a small garden to have a couple of different flowers blooming throughout the season and will not leave the garden bare of blooms. Following the suggested ratio in any size of garden will increase the potential for more pollinators, other beneficial insects, and plenty of blooms to harvest. The more blooming flowers are added throughout the seasons, the more benefits will be reaped.

The key to having an ongoing supply of blooms is to have two planting areas for flowers. One area or bed will be blooming while the other spot is in the process of being planted and growing for the next season of flowers. While enjoying the spring blooms, you will be planting the summer blooms, and so on. Keeping the flowers of the season clustered together leads to easier tending, but it also makes it easier for the pollinators and beneficial insects to find the flowers.

One of the challenges we face when planning our gardens is how to divvy up the available space to make the best use of it. This applies not just to the ratio of vegetables to flowers but also to achieving a balance within each of those groups. We almost always want to grow more than we have space for. You can expand and even double what you can grow in a space by realizing that annual plants have a lifespan. When they stop contributing to the effort of a working garden, it is time for them to be replaced.

An early-spring harvest of leaf lettuce, dill, and bachelor buttons gets spring off to a tasty and beautiful start.

Allowing space for two patches of flowers in the vegetable garden provides a consistent supply of blooms in the garden and for the kitchen table.

Learning and practicing this discipline will help you make the most of every square foot of the garden and keep maintenance to a minimum while maximizing your rewards. Removing a crop at its decline and replacing it with a fresh planting is my secret to reaping such abundance from my small urban farm. In my experience, the next natural step in the life of a tired plant is removal. There is no point in leaving a plant to decline until the end of the growing season. These removed plants continue to contribute to the garden in a whole new way; I toss them into the compost heap, where they will become next year's soil food (more on that in Chapter 8).

Treating this garden as a working garden and not a landscape bed will open the door for more blooms to harvest. I suggest that this garden, no matter how big or small, be located out of view from where you might work or sit in the house. Why? Because while this garden will be a beauty, its purpose is to produce beautiful blooms and delicious organic food for the table, not to be confused with and viewed as a landscape bed. Folks who can view and enjoy the garden from inside their home are less likely to harvest the flowers as often as needed and rotate crops as suggested because they enjoy looking at them so much. My suggestion is based on years of fielding questions and listening to experiences of gardeners. The perfect place to locate this garden is exactly where you don't landscape because it is out of sight, perhaps along the side of the house or behind the garage.

After I learned that I should plant sweet peas in the fall for my winter Hardiness Zone 7, I became an overnight success with them!

NATURALLY EXTEND THE SEASON
WITH SEASONAL PLANTINGS

A challenge I faced as a new gardener and then as a farmer was figuring out the proper time to plant specific plants. In my early days, I relied on other gardeners in my area for the cue to plant some of the more common flowers and vegetables, but when I ventured into seed starting, the selection of what I could grow exploded. Other gardeners could not tell me when the planting time was for most of these new plants. I naively thought all plants were to be planted in spring after the frost had passed, and it had not occurred to me that some flower

and vegetable plants could be planted earlier. I had yet to learn that some plants not only survived cool to cold conditions but actually benefited from them. I did not realize that some plants could even be planted more than once in a season to spread out the harvest.

Eventually my eyes were opened to the truth: there are different types of annuals, and each type has different planting times and preferred growing conditions. I learned that the workhorses of the cut-flower and vegetable garden are the cool-season hardy annuals and the warm-season tender annuals. Both groups include many flowers, vegetables, and herbs to choose from, and both types live for 1 year. The key difference is in the planting times.

For the purposes of this book, I have included and defined the following types of plants:

- **Warm-Season Tender Annuals:** Sometimes called summer annuals, these plants live for 1 year and prefer warm to hot temperatures; frost kills them. This family includes zinnias, sunflowers, tomatoes, basil, and others. They thrive in warm to hot conditions and prefer to be planted when the soil has warmed. Don't be fooled by warm daytime air temperatures alone—it's the nighttime temperatures that dictate the season. Don't let yourself plant until nighttime temperatures are staying above 60 degrees Fahrenheit. Warm-season tender annual seeds and plants resent cool conditions and will resist sprouting and thriving.

- **Cool-Season Hardy Annuals:** Also called winter annuals, these plants live for 1 year and prefer cool to cold temperatures; some can even survive below-freezing temperatures. This family includes plants such as pansies, sweet peas, snapdragons, spinach, kale, and others. They thrive in cool conditions and better tolerate growing in heat and humidity if they've become established in cool conditions. They can be planted as early as 6 to 8 weeks *before* the last expected spring frost everywhere. In some areas, they can be planted in fall to winter over. A midsummer planting of hardy annuals can provide an abundance of blooms right up to the first hard frost. While they may not be very happy with the warm conditions at the planting time, they grow into their preferred cool conditions as the season moves along.

- **Perennials:** These plants typically live for more than 2 years; the foliage dies back in the winter and regrows in spring. This family includes peonies, mint, sedums, and others. The best time to plant perennials is in late summer and fall so they have time to become established before winter weather begins. This gives the benefit of spending the winter settling in and growing roots before facing their first season of producing flowers or fruits.

Planting fall-colored tender annuals in midsummer will provide beautiful colors of the season. Remember to harvest flowers like hydrangeas from the surrounding landscape.

Including warm-season and cool-season annuals in the garden allows you to naturally extend the season by starting earlier in spring and going later into fall. It also opens the door to one of the foundation rules of organic gardening: planting any given plant at its proper time empowers the plant to become established and thrive with as little help from the gardener as possible. There is no question in my experience that plants planted in "their time" grow up to be more disease resistant, more drought tolerant, and more productive. Once you become familiar with the conditions each type of annual prefers and when those conditions occur in your garden, it becomes second nature to know when a given plant prefers to be planted. When we mimic and follow nature's example, it is so much easier and more fun to garden. If I don't know which type of annual a plant is, I enter the name into an Internet search engine and ask that question.

SUCCESSION PLANTING

As I was gaining experience through those first years of vegetable gardening, I never gave much thought to having a reliable, steady flow of vegetables from the garden; I just took the harvest the garden offered as it came. While it could be overwhelming for a vegetable crop to come in all at one time, I either preserved what we couldn't eat fresh or gave it away. It was becoming a flower farmer that brought me face to face with how to make the most of and to stretch out each season.

As my cut-flower farm began to grow, my commercial customers wanted to purchase my flowers as early and late in the growing season as possible and to have the same flowers available week to week for as long as

possible. By including cool-season annuals in the garden, I was able to have flowers earlier and again later in the growing season, but it was stretching those warm-season harvests out over time that I needed to improve on. This need pushed me to learn about succession planting.

Succession planting is planting smaller numbers of the same plant at intervals in place of one large planting. This practice is coupled with removing a crop as soon as it is done so the space can be replanted immediately. This not only extends the harvest over a longer period time but also makes the most of the garden space. The staggered plantings will bloom and are harvested over a period of time instead of all blooming at once. I apply these practices to many vegetable plantings as well as flowers. Here are some examples I have used with success.

SEVERAL PLANTINGS OF A ONETIME HARVEST PLANT

Most cut-flower-variety sunflowers are single stemmed, one bloom per plant. To have a continuous weekly harvest of sunflower blooms from spring right up to the first fall frost, we plant a set number of 2-week-old sunflower transplants each week for as many weeks as possible. This provides a steady weekly harvest. For many years, we planted 1,200 plants a week from spring until late summer. After each weekly bed is harvested, the bed is immediately cleared and prepared for another planting. A weekly or biweekly planting of 10 to 15 sunflowers will keep the home gardener in flowers. In the vegetable world, the radish is an example of a one-harvest-per-plant crop. Planting a short row of seeds each week can provide a weekly harvest of radishes throughout the salad season.

A WARM-SEASON TENDER ANNUAL RECIPE

Planting some of the workhorse summer annuals more than once in a season can ensure a steady supply of flowers. This practice provides a backup in the event of a loss from weather or pests. This is also how to have an abundant harvest in late summer to fall. The garden space available and the length of your warm-growing season will dictate how many plantings you can make.

In my Mid-Atlantic Zone 7 garden, I plant four repetitive summer flower recipes of annual plants. The first round of transplants goes in after the last expected spring frost date, and the rest are spaced once a month

Planting a few sunflowers each week will provide nonstop sunflowers all season. The 'ProCut' series will bloom into fall as the days get shorter.

for a total of four plantings. This keeps my garden producing abundance and quality as long as possible. The flower recipe includes annuals such as several colors and varieties of zinnias, cockscomb, celosia, ageratum, and basil for foliage. Oftentimes a "recipe" planting will go into the garden space following a sunflower harvest. Pulling and planting becomes the norm and will keep the garden packed with plants producing abundance. This same repetitive planting can also be used for vegetables such as tomatoes, squash, peppers, and more.

To find the latest planting date for a specific plant in your garden, you must first know the average days to maturity for that plant and your first expected fall frost. Count back from that first expected frost date the maturity days needed for the plant to grow and produce and add days to reap the harvest. These additional plantings are the secret to an abundant fall harvest of summer flowers and vegetables.

STRETCHING COOL-SEASON HARDY ANNUALS

Cool-season hardy annuals may have a narrower planting window depending on where you garden, but they too can be manipulated to extend their season with more than one planting. Those regions that can plant cool-season hardy annuals in fall to winter over

can repeat that planting in early spring to extend the harvest. In this case, the second early-spring planting may not be as robust or productive as the fall planting, but it is still worth doing. In gardens with long, harsh winters that prohibit fall plantings, one planting in early spring and a second planting a few weeks later will help extend the harvest season. In regions that have a short warm growing season, gardeners will likely be able to keep the cool-season hardy annuals producing longer into summer than those with longer warm seasons. See Chapter 5 for more on cool-season hardy annuals.

THE GARDEN PLANS

In the appendix on page 164, I offer two plans for vegetable gardens with flowers. One is a city salad garden, and the other is a larger plot garden. The examples show how a garden's plantings can evolve through the growing seasons with succession plantings, making the most of the given space. While seasons occur on different timetables for various regions, these examples are intended to give seasonal ideas that may be adapted to your garden's timing.

The Cutting Garden

Keeping a cutting garden in tiptop shape means it must be set up to do its job. That process starts with the selection of plants to grow, moves on to preventing the tall flowers from falling over, and ends with the best part—harvesting.

It may be frightening at first to harvest as I recommend, but be assured: your loyalty to the suggested harvesting habits will reap a reward. The more often you harvest, the more flowers will be produced.

The secret of a cutting garden is perhaps the sweetest part. Because of the abundance of long-lasting cut flowers from a little patch, there will be plenty of flowers to share with family and friends.

OPPOSITE: *A summer harvest of zinnia 'Oklahoma' and 'Benary's Giant', cosmos 'Double Click', sunflower 'ProCut', celosia 'Sunday', tansy, and black-eyed Susan 'Triloba'.*

Beri and I are heading out to harvest early in the season. 'Benary's Giant' zinnias will grow over 36 inches tall, making them excellent cut flowers.

GROW LIKE A PRO

Taking a few specific steps in the beginning will lead to an easy-keeper cut-flower garden that will keep on going throughout the growing season.

IT'S ALL ABOUT VARIETY

I give space in the cutting garden only to proven cut-flower variety plants. In general, this means that the plant variety produces prolific blooms when harvested, has long stems with stiff necks, and holds up looking good in a vase for at least 7 days. In catalogs, you will often see the scissor emblem on seeds and plants that indicates it is a good cut flower. All the flower varieties suggested in this book are excellent cut flowers that I have grown. If I come across an irresistible flower but am not sure how it will hold up as a cut flower, it goes in another garden space for its first growing and harvesting season. Only when it proves worthy does it warrant space in the working cutting garden.

Don't be fooled by the look of a flower when making variety selections for the cutting garden. The blooms within a genus can appear the same, but the characteristics between varieties can differ. Take the zinnia, for example. There are several varieties with large, beautiful flowers, but their height varies. Some grow 18 inches tall, and others grow 48 inches tall. The 18-inch plants would be a sad disappointment in a cutting garden. I know because I spoke to a gardener who mistakenly purchased plants based on the bloom alone. All the hard work and anticipation were snuffed out by simply selecting the wrong plant.

PLANT SPACING

The thing about a cutting garden is that once it starts to bloom, it must be harvested nonstop. To a plant, harvesting is like receiving a hard pruning a couple of times a week. This is why it is possible to space plants closer

continued on page 46

To accurately space plants, we use the 6-by-6-inch square flower support netting that will ultimately be installed to support the tall flower stems.

Sunflower Facts

Why pollenless sunflowers for cut flowers?

Because they simply are better suited to the vase. Sunflowers that produce pollen can stir up allergies, drop the yellow pollen onto tabletops, and contribute to a shortened vase life. The pollenless varieties do not litter tabletops, have the stiffest straight necks, and typically last up to 14 days in a vase.

Pollenless sunflowers produce abundant nectar!

Even though pollenless sunflowers do not produce pollen, they do produce abundant nectar. Dr. John Dole of North Carolina State University, a leading cut-flower industry researcher and teacher of floriculture, confirms that pollenless sunflowers do produce nectar. Pollen and nectar plants are important to attracting pollinators. Some, such as butterflies, are primarily there for the sugar in the nectar. A cut-flower garden grown in a diverse habitat will work together with other native trees and shrubs to provide a rich pollen and nectar presence.

I prefer to grow the single-stemmed pollenless plants for cut flowers. They produce a strong stem with the classic sunflower bloom, and they do it quickly. The sunflower genus is full of many beautiful choices and varying characteristics. In addition to single-stem and branching types, there are different colors and styles of blooms. Both types offer pollenless varieties.

Determine bloom size by the spacing. One of the reasons I prefer to grow a single-stem pollenless sunflower variety is for the ease of manipulating the size of the bloom. It's up to you whether they will grow big or small, as the distance between plants determines the size of the bloom. To grow what I consider the perfect 4- to 5-inch bouquet sunflower, plant four rows to a 36-inch-wide bed and space the plants 6 inches apart in the row. If you spaced the same variety of sunflowers 12 inches apart, they would be twice as large.

Want a whopper sunflower?

If you want to grow for the fair or just want a whopper in the garden, you can do it. First select a giant-variety sunflower seed, such as 'Mammoth'. I have grown it to well over 14 feet in my garden, and birds enjoy the huge heads.

1. Before planting, add organic dry fertilizer to the planting area according to instructions. Sunflowers are very hungry flowers.

2. Start the seed indoors and transplant the 2-week-old plants into the garden.

3. Space the transplants 2 feet apart in all directions.

4. Water the baby plants daily for the first week and then once a week.

5. Add organic liquid fertilizer to the watering can each week to give the growing plants a good meal.

6. Troubles with rabbits or other varmints eating plants? Cover the plants with a floating row cover for the first 10 to 14 days. See more on this on page 141.

7. Place a sturdy 10-foot-long, 2-inch galvanized pipe next to each seedling, driving it into the soil at least 1 foot. Don't be fooled by that little plant. It grows into a big, beautiful flower head with a 3-inch trunk that will topple in wind and rain. I use self-gripping plant tape to secure the sunflower to the stake as it grows.

While this huge sunflower is not a good cut flower, it is totally fun! The birds love eating the seeds from the mature head and bathing in the little pool of water that gathers in the depression on the backside of the aging bloom.

together than in a landscape setting. For spacing in my garden, I tend to err on the side of placing plants closer together rather than farther apart unless the plants show signs of pressure, such as disease or pests.

When cut-flower plant spacing is combined with other practices in this book, the garden produces tremendous abundance. When starting a new garden space, or if there are difficult conditions to overcome in an existing bed, I would recommend erring on the side of wider spacing. In a 30- to 36-inch-wide raised bed, I plant four rows of most annuals mentioned in this book. The rows are spaced equally apart across the width of the bed. In the row, plants are spaced 6 to 12 inches apart. Most often I follow the 6-inch spacing. For specific flower-spacing recommendations, see Chapters 4 and 5.

SUPPORT THE STEMS

Cut-flower-variety plants grow tall. This characteristic, coupled with potentially heavy flower heads, makes the stems vulnerable to toppling over in the garden. Even those plants that don't grow so tall can still get bent in the garden, resulting in unusable stems. I can't express how disappointing this is to discover.

Installing flower support early in the life of the cutting garden can prevent crooked stems and prevent heavy flower heads from going down in the garden. While there are several methods of support available, I have found in our windy and frequently rainy setting that the plastic flower-support netting with sturdy stakes works best. It is easily installed, useable for more than one season, and provides great support to even the most abundant canopy of flowers.

Here are a few tips:

- It's best to install support netting while the plants are under 12 inches tall. This allows the plants to naturally grow up through the netting. It will practically disappear from sight as the plants mature.

- I prefer to use either metal T-posts or 2-inch-square wooden garden stakes. Install a stake every 6 feet on both sides of the bed.

- Stakes are easiest to pound into the ground with a sledgehammer.

- Netting is easily cut with scissors and should rest on the stakes at half of the mature height of the

Flower-support netting will prevent flowers from going over in rain and wind. Install when plants are young and they will easily grow up through the netting.

planting. Make sure it's pulled taut to keep it at the desired height. For a mixed cutting garden, a 20-inch netting height works well.

HARVESTING HABITS

The success of the cutting garden will depend on your harvesting habits. In fact, its very life depends on them. This simple truth goes for vegetable and herb gardens as well. The garden, no matter how big or small, must be harvested on a regular basis. Sadly, a common occurrence is to harvest only when a need arises, which leads the garden to its untimely end.

The potential of a cutting garden can only be realized if the harvesting is done when and where the plant needs it. This encourages the plants to continue producing new growth throughout the season, and new growth leads to more flowers. This continuous act of cutting keeps annual plants on their toes producing fresh foliage and flowers and in the very best of health.

The frequency of the recommended harvesting practice is actually the opposite of what most folks imagine. This misconception has led many a gardener astray and is responsible for the common demise of cut-flower gardening dreams. I find gardeners either are reluctant to cut for fear that more flowers will not grow or just can't bear to cut the flowers because they are enjoying how beautiful they look in the garden viewed from indoors. Make sure you follow a cutting-garden setup from the get-go to avoid these traps!

Cut-Flower-Gardening Ground Rules

1. Freestanding beds that are accessible from all sides are the easiest to tend and harvest over time. I prefer beds 36 inches wide or less so I can comfortably reach the middle without putting my foot on the bed.

2. Variety does matter. Sticking with cut-flower varieties ensures a season of flowers that hold up in the vase.

3. Plant spacing is closer in a cutting garden. Plants are continually harvested once they start blooming. This equals a serious pruning one to two times a week.

4. Provide support for the tall stems before the plants reach 12 inches tall.

5. Get familiar with the flowers you are growing. Know at what stage of opening the bloom should be harvested, where to make the first cut to set the plant up to produce all season, and where to make the subsequent harvesting cuts.

6. Use plastic containers for harvesting that are easy to clean. For gardens that provide a few handfuls of flowers each week, containers with narrow openings are best.

7. Garden flowers appreciate cut-flower food in the harvest bucket and vase. It keeps the stem developing, helps keep the water clean, and prevents clogged stems.

8. Find unlikely containers to use as vases that are shorter and have a wide opening.

9. Make a harvesting date with your cutting garden each week and stick to it.

10. Share the bounty!

HOW OFTEN TO HARVEST?

Harvesting flowers twice a week is a good habit to establish. On my farm, Monday and Thursday are the harvest days, and it happens like clockwork. Make a date with the cutting garden and stick to it. It's a fun habit to establish and won't take long if you stick to the job at hand: only harvesting. It is very easy to get distracted in other directions, such as pulling a few weeds or checking up on the tomato plants. Harvesting is a job that needs to be completed on time to keep the garden in good shape. Finish harvesting and then go back to other chores if needed.

Frequent harvesting also preserves the quality of the blooms. Flowers that sit out in the garden for an extended period of time are at risk of being damaged by insects and weather. Timely harvesting minimizes the chances of that happening. The more often you harvest, the better the quality of the blooms will be.

WHERE TO MAKE THE CUT?

When harvesting, a natural tendency is to *not* make the harvesting cut as deep on the plant as it needs to be. This is especially true for that first cut on an annual plant, which establishes the branching of the plant for

On harvest day, the fuller, more developed flowers will be cut. Those left behind—the understory less developed—will be ready on the next harvest day.

the entire season. The next sprout of growth on a plant will come from where a cut is made on the plant. The strongest and sturdiest stems grow from near the bottom of the plant, so by making the first cut deep in the plant, the plant will branch and produce more stems of better quality throughout the season.

Harvesting the blooms on a regular basis keeps the plant growing new shoots. This constant cutting keeps the plant healthy and producing.

The first cut on an annual plant with a central stem (pictured on the opposite page) should be made at almost ground level, just above three to four side shoots. All the subsequent harvest cuts are made at or near the end of the stem being cut. Harvesting with flower-support netting installed will be frustrating when you are first learning how to do it, especially for that first cut. Without the netting, however, there will be few straight stems to harvest! The subsequent cuts are far easier to manage, and it goes more smoothly as you become familiar. Do your best to not pull the flower-support netting up and off the bed when harvesting. Sometimes it is easier to pull that multibranched first cut down under the netting versus pulling it up through the netting, but keeping the netting in place beyond the first cut makes the rest of the season so much easier!

COMMON PITFALLS
Avoid these harvesting mistakes:

1. **Making the first cut too high on the plant.** This mistake plagues the plant with troubles all season. The regrowth will be short and vulnerable to breaking. The plant quickly becomes top heavy with all the sprouting going on in the canopy instead of at the base of the plant. The base of the plant also becomes old and worn, whereas it could have been popping with new growth.

2. **Not stripping enough foliage from the cut stem**. Every piece of foliage left on the stem takes a toll on the vase life of the flower. Removing all that does not contribute to its end use can add days of life to the cut flower.

3. **Harvesting at the wrong time of day.** The best time to harvest is either before or after the heat

The First Cut: *To establish branching on zinnias, celosias, basils, and other central-stem plants, the cut must be made low on the plant as pictured.*

The Next Cuts: *The branching stems should be cut at the base of their stem. The deeper the harvest cuts are made, the stronger stems will regrow.*

of the day. Directly after harvesting, place the harvested flowers in a cool spot for at least 4 hours to recover. (Overnight is preferred.)

4. **Using a bucket with a wide opening for just a few flowers.** If you are expecting to cut a couple of handfuls of flowers, use a plastic container with a narrower neck. This does the best job of keeping the stems upright and not allowing them to slide down the side of the bucket. Stems that get crooked in the bucket seldom return to straight.

5. **Jumbling various stem lengths together in one bucket.** To prevent crushing blooms in the harvest container, group similar lengths together. My aim is to cut stems to similar lengths so the blooms are shoulder to shoulder. This allows any bloom moisture to dry and prevents smooshed blossoms. When harvesting flowers that are shorter, such as sweet peas, I take a special container to the garden just for them.

6. **Leaving the harvest container sitting in the sun.** Setting the harvest container in the shade

is always the first choice, but it's not always practical. Having it in the garden close at hand is the most convenient, but that usually means it stays in full sun. That is okay if you harvest and then move it indoors as soon as possible.

PEONY PICKING

Peonies are ready to harvest when the buds go from feeling hard like a marble to soft like a marshmallow. As peony buds begin to swell but before they begin to open, I patrol the peony patch twice a day. I stop to squeeze each bud that looks like it might be ready. I cut the stem when the bud feels like a marshmallow, always leaving one-third of the stem and foliage on the plant. The buds will open indoors out of the hot sun and drying winds, which makes for longer-lasting, more pristine blooms.

Harvesting sunflowers when the petals lift off of the center as pictured will prevent damage to the tender petals by grasshoppers and beetles. They will continue to open indoors.

Harvesting gear makes the job more efficient and will result in longer-lasting flowers. Wear gloves to strip foliage or your hands will be stained.

WHICH BLOOMS TO CUT?

A flower's stage of opening when it is harvested plays a role in the quality of the bloom and how long it will last in a vase. Some flowers continue to open after they are cut, while others stop dead in their tracks the minute the cut is made. Knowing how the flowers you are growing react will dictate what flowers to cut on each harvest day. I have included the best stage to harvest for the flowers featured in this book in Chapters 4 and 5.

HARVESTING GEAR

For a quick and successful harvest, I take these essential items with me to the garden: clean plastic harvest containers, water treated with cut-flower pretreatments or flower food, lightweight bypass shears and pouch, and breathable gloves that fit like a second skin.

- Harvest buckets are treated like glasses to drink water from; they are washed after each use with dishwashing liquid and allowed to dry. Using clean containers is the number-one thing to do to extend flowers' vase life.

- Use a measuring stick (I mark mine at ½ and 1 gallon of water for my containers) to fill the buckets with water and then add the appropriate amount of pretreatment or flower food to go to the garden for harvest.

- Bypass shears make the cleanest harvesting cut. How do I keep up with my shears for more than a season? I always wear a pouch that slips onto my pant pocket so I won't lay my shears down in the garden.

- I do not go to the garden without gloves. Stripping foliage from stems will destroy your hands. Some foliage will also stain your hands beyond what soap can remove. It's not a good long-term idea to harvest and strip without gloves.

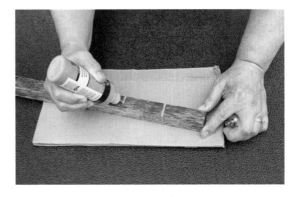

To make a measuring stick: Fill the container with water. Dip the stick in to make a water-level mark. Using craft puffy paint, mark the fill-to line.

CUT-FLOWER FOOD

We use pretreatments and cut-flower foods because they help our flowers last longer and look better. The way homegrown garden flowers are handled predisposes them to drink a lot of water. Store-bought flowers have traveled thousands of miles in a box, and their clogged stems prevent drinking. Because these traveling stems are bred to sustain life without water for long periods of time, we might not even realize that their stems are clogged and they aren't drinking, except for the fact that their vase rarely needs more water. Since they are not drinking, it looks like the flower food added to the vase is unused, which leads us to believe flower food doesn't make a difference in the life of a flower stem. It doesn't help flowers that aren't drinking, but this is not the case with flowers cut from your garden—they drink and will benefit from what you put in the water. Always follow package directions and measure appropriately; under- or overdosing does not benefit the flowers.

PRETREATMENTS

I use two types of pretreatments. One is an instant hydrating solution to help unclog flower stems, and the other is a time-released chlorine tablet that delays the development of bacteria in the water and extends the vase life. Pretreatments are not a replacement for cut-flower food but something to be done when you want to restore a droopy head or extend vase life.

The symptoms of clogged stems are droopy heads or an unchanged water level in the vase. I cut off 2 inches of stem from all store-bought flowers and then dip them into instant hydrating solution, according to package directions. I try this also on garden flowers that may still be droopy the morning after harvest. This treatment reopens the vessels in the stem and allows water to flow up the stem. Following this treatment, the flowers should be immediately placed into water with the chlorine-tab pretreatment to give a deep drink of clean, bacteria-free water.

On the farm, a chlorine tablet goes into every harvest bucket to delay bacteria development. I believe that every step in the harvesting practice can add a little more time to the vase life of a flower. The flowers sit in the treatment for at least 4 hours but can stay in it for up to 72 hours. From the pretreatment containers, the flowers go into a delivery container or a vase with fresh cut-flower food.

IN THE VASE

Adding cut-flower food to the vase has a profound effect on the flowers. The foliage stays green, and the buds continue to mature. It prevents bacteria in the water and

Harvest Troubleshooting

1. **Drooping heads never recover after harvesting.** If flowers wilt during or right after harvesting and are still wilted the following morning, the next time I harvest, I consider either stripping more foliage from the stem or harvesting first thing in the morning before heat.

2. **Flowers are wilting prematurely.** This may seem obvious, but check the vase for water. Garden flowers deplete vases of water faster than store-bought flowers. In the event that the vase runs dry, cut 1 to 2 inches off the stems and refill the vase. Select vases that hold a lot of water.

3. **Flowers don't last very long in the vase.** Some flowers simply don't last in the vase no matter what you do—is it a cut-flower variety? Sometimes, harvesting stems of cut-flower-variety plants too early in their development will lead to wilting that cannot be overcome.

4. **The vase water is cloudy.** There is a group of flowers known as the Dirty Dozen whose members pollute the container water more than other flowers by oozing excessive debris from their stems into the water. This debris feeds bacteria, which in turn makes the water cloudy and leads to a shortened vase life. Some of the offenders include zinnias, marigolds, snapdragons, and sunflowers. Using a pretreatment followed by flower food delays this cycle and can help flowers last longer in the vase.

Making a Simple Bouquet

All garden flowers are beautiful and can be included in a bouquet together. What really helped me become a more confident bouquet maker was realizing that the problem wasn't my bouquet making at all but the vase I was putting the flowers into. When I began using some of the items in my cabinet that weren't intended as flower vases at all, it got easier, and my bouquet-making skills soared.

I went from using tall, narrow flower vases that strangle bouquets to shorter containers with wide openings, such as a gravy boat and my grandma's little white McCoy pitcher. I also found cutting the bouquets much shorter than I ever imagined allowed me to use them in more personal areas of my home. Placing a short but wide bouquet on my kitchen table worked because it didn't block the view of the person sitting across from me. Bedside and coffee tables can easily host simple, short bouquets. Going short also highlights the best part of the bouquet—the canopy of blooms—displaying them just below eye level to be enjoyed by those seated and standing.

I follow a few simple steps whether I am making a little handheld bouquet for my kitchen table or 200 mixed bouquets for the supermarket:

- Allow the flowers to recover from harvesting for at least 4 hours before you start the bouquet-making process of pulling them in and out of water.

- Remove stem foliage that falls below the vase water level and is not contributing to the bouquet. (I do this in the garden as I harvest.)

- Organize flowers that are alike together in the bucket or lay them out sorted on the table for a short period of time while making the bouquet.

- Group three stems of smaller flowers together for more impact.

- Have the container ready and waiting with water.

ABOVE: *Foliage should be removed from the stems, leaving only the top few inches. Group similar flowers together on the tabletop.*

ABOVE: *Holding the lowest third of the stem, arrange the foliage and filler stems in a pleasing way. This builds the structure of the bouquet.*

ABOVE: *Place the flowers one by one, using up an entire group of similar flowers, adding from the side and also poking down through the foliage.*

LEFT: *Continue adding flowers, turning the bouquet in hand as you go. Keep the flowerheads fairly level, with those on the outer edges a little lower.*

RIGHT: *With the vase close to the edge, hold the bouquet next to it with the lowest blooms at the vase rim to see where to make the cut.*

Check the vase daily for water and refill as needed. A common reason for the demise of garden flowers is the vase going dry.

also maintains the pH level to prevent water bubbles that can clog stems. The big difference between homemade remedies and ready-to-use flower conditioning products is their stability over a period of time. Most homemade recipes should be changed out almost daily because they lose their effectiveness so quickly.

I recommend refreshing the vase water after 3 to 4 days. Lift the bouquet from the container, holding it together throughout the refreshing process. Cut 1 inch off the stems, and rinse the stems off by holding the ends under running water to wash away any residue. Rinse the vase and refill it with water. The flowers are now ready to go for a few more days. Since the bouquet is in the second half of its vase life by this time, I usually do not add flower food to the new water. Check the water level in the vase daily and refresh as needed.

THE GREATEST GIFT OF THE GARDEN

Sharing bouquets with family and friends is one of the greatest gifts the garden can give. Even the smallest cutting garden will give an abundance of weekly long-lasting cut flowers. If you resist the urge to share the blooms, you may find your home taking on the look of a funeral parlor by the height of the growing season! Or, worse, if you stop harvesting as recommended because you have too many, your cutting garden will begin its premature spiral to the end.

Some of the most unlikely recipients will be smitten your flowers. I keep a year-round running list of those I'd like to surprise with a bouquet. During the growing season, when I am running on empty most days from working outdoors in the heat, it is so easy to look at the list to be reminded of someone I may not have thought of or seen for months. I drop the rubber-banded bouquet into an inexpensive plastic bucket with a short note saying "Thinking of you" or "Your works are appreciated" and leave it on the unsuspecting receiver's porch. I don't use vases because they are too easy to break. I don't even knock normally—it's having the ability to make this practice quick and easy that allows me to exercise it so much. Grow the love of flowers in others and share the bounty.

One of the joys of flowers is sharing them. Drop a bouquet on the porch of someone who needs a lift, and you will be lifted too.

SECTION 2

Plants by Season

The annual plant leads a brief but miraculous life. It all starts with a little seed. The seed sprouts and grows quickly. Before you know it, flower buds are coming along, and the blooming begins. As soon as the flowers begin to fade, the plant takes on its real mission in life—making seed for the next generation of plants.

But if the flowers are cut in their prime and don't get a chance to fade, guess what? They won't make seed. And so the annual plant must keep on making new flowers. No other garden plant has the annual's ability to keep putting out new blooms for as long as we keep harvesting them. What wonderful news for gardeners!

Annuals never give up trying to complete their life's work. The plants keep sending out shoots to make new flowers in an effort to make seed, and we keep cutting them off. This is the secret to keeping the blooms coming—tricking that plant into keeping on trying to make seed. The more you harvest annual plants (flowers, vegetables, and herbs), the more shoots will come. This results in an abundance of flowers. It's simple as long as you keep harvesting!

OPPOSITE: *An early-summer bouquet: sunflower 'ProCut', zinnia 'Benary's Giant', millet 'Lime Light Spray', celosia 'Spring Green', snapdragon 'Opus', pincushions 'Blue Cockade' and 'Fata Morgana', bupleurum, false Queen Anne's lace, yarrow 'Colorado Sunset', dill 'Bouquet', and feverfew 'Common'.*

LEFT: *Warm-season tender annuals like the sunflower thrive in heat. They resent being planted and grown in cool conditions. Plant once nighttime temperatures are above 60 degrees.*

BELOW: *Cool-season hardy annuals like poppies thrive in cool to cold conditions. Most gardeners can grow this family successfully, even in the South, when they are planted at the proper time.*

SHORT-LIVED VERSUS LONG-LIVED PLANTS

I am a huge advocate of annuals. I love that all things can be made new each season. And then there is the whole delightful process, from seed shopping to harvesting armloads of flowers—I enjoy every step. Let's face it: if you like the act of selecting, planting, and all that comes along with it, annuals may be for you too. It's a joy to watch seeds sprout, plants grow, buds swell, and flowers bloom.

In my first few years of gardening, I had yet to appreciate the annual and its short life. But after I spent those first years filling my gardens with perennials and bulbs that returned year after year, I soon learned I was running out of garden space!

So I slowed down on permanent planting and began ramping up on annual plantings. Growing annuals means you have a garden area that offers a clean slate each season. You can experiment with what to plant because if it doesn't work out, well, there's always next season to try something else! Annuals are a short-term commitment, a quality many seem to appreciate, especially in a garden.

At first I thought my annual garden was going to be more work than my permanent plantings filled with

perennials, but what I discovered was that maintaining perennials has its own set of ongoing chores. They need to be kept weed free year round. Many need to be lifted and split every 3 or 4 years to stay healthy and keep producing blooms. While certainly worth the work, it's not the carefree gardening many think perennials represent. Grouping plant families like these together with their similar needs goes a long way in streamlining maintenance and providing opportunities in a garden. One of the best things about areas that include perennials and bulbs is that they offer year-round habitat for all the beneficial creatures that flowers invite to the garden.

THE CONFUSING ANNUAL

There is often confusion surrounding the term *annual plant*. Most folks automatically think of well-known warm-season tender annuals such as marigolds, petunias, and tomatoes. There is another family of annuals that is not well known: the cool-season hardy annuals such as pansies, snapdragons, and kale. These two annual families share the same life cycle but are planted at very different times. Together they are the workhorses of cut-flower and vegetable gardens, allowing the gardener to start earlier and go later in the season.

- **Warm-season tender annuals** thrive in warm to hot weather. Plant them in the garden after all danger of frost has passed and the soil has warmed. The plants are killed by the first fall frost.

- **Cool-season hardy annuals** thrive in cool to cold weather. Plant them in the garden in early spring or in the fall in milder winter regions. The plants begin to decline as hot, humid weather begins.

ANNUALS FROM SEED

Learning to start annuals from seed opens the door to endless variety selection. No longer are you restricted to the plants your local nursery is offering; the only limitation will be your ability to troll countless seed catalogs. Starting your own organic transplants allows you to have the plants available at the right time for your garden. The transplant will typically be healthier when homegrown as well.

Annuals are easy to start from seed. A set of 20 ¾-inch mini-soil blocks with 3-week-old ageratum transplants ready to be planted.

Starting a garden from seed is far more economical than purchasing plants. In addition, starting and growing your own plants for the garden can be a source of incredible satisfaction. Once you understand just a few seed-starting ground rules, the mysterious seed will no longer be quite so mysterious. For my seed-starting ground rules, see Chapter 8.

SEED STARTING

For each featured flower in this section, I list how the seed prefers to be started:

1. Indoors

2. Outdoors in the garden

3. Indoors or outdoors (can be started either way)

For a flower that has both methods listed (#3 above), the method listed first is the one I prefer. "Start indoors or plant in the garden" means it can be started either way, but I find starting indoors more successful. Instructions are included for starting seeds indoors using the English soil-blocking method or in a container. See more on seed starting in Chapter 8.

Finding where annuals fit in your garden will start you down a path you will not regret taking. Each new season will be full of new choices and opportunities, and it all begins with a seed.

Warm-Season Tender Annuals

As the temperatures of summer begin to rise, warm-season tender annuals start coming into their own. With their hot, beautiful colors, they bring to the garden the jewels of the season—butterflies, hummingbirds, and bees of all kinds. These plants will start blooming in late spring and keep on until frost, assuming you are harvesting the blooms regularly. That first planting in spring can still be pumping out the blooms in fall if they are kept weeded, watered, and fed.

OPPOSITE: *I love adorning our well lid located in my shade garden with flowers for special occasions—zinnia 'Benary's Giant'.*

WHEN SHOULD
I PLANT?

This family of plants likes to start life in the garden when the soil and air temperatures are warm to hot. Once the last expected spring frost date has passed and nighttime temperatures are staying consistently above 60 degrees Fahrenheit with summer drawing near, it is time to plant. A common mistake is to plant too early, when the soil and air may still be chilly. Plants that survive this blunder will get off to a slower start and can suffer consequences the rest of the season. The best opportunity for plants to grow a deep and healthy root system is when they're planted into their preferred planting time. They love warm soil and air temperatures and will reward you when that is given. Once well established, these plants can face the heat and humidity of summer with little intervention.

TRANSPLANTS
OR SEEDS?

Transplants to the garden will start blooming weeks earlier than seeds planted directly in the garden. This is a result of starting the seeds indoors and having plants

ABOVE AND LEFT: *There are many choices when it comes to starting seeds indoors. Select from a container such as this 128-size plug tray or soil blocks.*

ready to go to the garden as soon as it is warm enough. Beyond earlier blooms, transplants can be planted into mulched beds that will have little to no weed pressure, and planting them at the suggested spacing means there's no need to thin. Most of the tending is done indoors before planting.

Seeds planted directly in the garden will start blooming weeks later than transplants due to all the growing time taking place in the garden after it is warm enough. In addition, seeds should be planted a little later than transplants so the soil has a chance to warm up even more for the strongest sprouting. Other challenges include preventing weeds from growing while waiting for the flower seeds to sprout and grow, thinning flower seedlings to the recommended spacing, and coping with insects and varmint pressure on young, vulnerable seedlings. All the tending is done outdoors in the garden.

TENDER-ANNUAL
PLANTING NOTES

Plants can look better in the garden and give more blooms over a longer period of time if you follow a couple of simple practices. Most often if the plants are left to their own accord they bloom and fizzle out after a few short weeks.

- Succession plant for an abundance of blooms in late summer and fall. Even gardeners with a shorter warm growing season can do this with quick crops such as zinnias, sunflowers, cosmos, and basil.

- In my long growing season, I plant a tender-annual recipe of plants three to four times over the course of summer. While I plant some of the same flowers over and over, I change the colors up to match the season, including bright, hot colors to bloom in summer and warm-color blooms for fall.

- Tender annuals should be harvested twice a week for the most blemish-free blooms and to keep the plant producing new shoots.

Favorite Tender-Annual Recipes

LEMON-LIME JOY

- **Zinnia:** 'Benary's Giant Bright Pink', 'Benary's Lime Green'

- **Zinnia:** 'Oklahoma Pink'

- **Millet:** 'Lime Light Spray'

- **Celosia (Cockscomb):** 'Cramer's Lemon Lime'

- **Basil:** 'Mrs. Burns' Lemon'

The Lemon-Lime Joy Bouquet. The lime-green flowers coupled with the fragrance of lemon basil makes this bouquet a knockout!

THE STAPLES

- **Zinnia:** 'Benary's Giant Mix'

- **Zinnia:** 'Oklahoma Mix'

- **Celosia (Cockscomb):** 'Chief Mix'

- **Celosia (Plume):** 'Pampas Plume'

- **Sunflower:** 'ProCut Orange'

- **Basil:** 'Cinnamon'

- **Basil:** 'Mrs. Burns' Lemon'

The Staples Bouquet. Using a short and wide vase—a gravy boat here—and cutting the stems short make this an excellent dinner table bouquet.

HOT SALSA MIX

- **Zinnia:** 'Benary's Giant' in scarlet, orange, and yellow

- **Celosia (Plume):** 'Forest Fire'

- **Celosia (Plume):** 'Sunday Orange'

- **Basil:** 'Mrs. Burns' Lemon'

- **Sunflower:** 'ProCut Orange'

Hot Salsa Mix Bouquet. A whole party theme can be planned around this mix of hot summer colors. A midsummer or fall favorite!

THE VICTORIAN

- **Zinnia:** 'Queen Red Lime', 'Queen Lime Green'

- **Zinnia:** 'Oklahoma Mix'

- **Celosia (Cockscomb):** Cramer's series mix

- **Celosia (Plume):** 'Sylphid'

- **Basil:** 'Cinnamon'

The Victorian Bouquet. As fall approaches, bouquets of more subtle colors fit in with the change of the seasons.

ZINNIAS, *Zinnia elegans*

AT A GLANCE

Sowing seed: Start indoors or in the garden

Sun: 6 or more hours

Height: 24" to 48"

Rows per 36" bed: 3 to 4

Spacing in the row: 6" to 12"

Days to bloom: 75 to 90 days

Deer resistance: Excellent

Container use: Excellent

Visitors to expect: Butterflies, bees, and other pollinators, hummingbirds, songbirds, and beneficial insects

The frilly petals and colors of the 'Giant Cactus' mix make them unique. They add texture to bouquets and to the garden.

'Peppermint Stick' blooms are 2 inches across and are like snowflakes—no two are alike! This flower is a favorite of butterflies and bees.

LISA'S GROUND RULES

Seed-Starting Tips

Zinnia seeds can be started indoors for an early start or planted out in the garden. For soil blocking, use the ¾-inch blocker, sticking the pointy end of the seed into the block until just the tail of the seed is sticking out. For starting in a small container or in the garden, cover the seed with ¼ inch of soil. Zinnia seeds need darkness to sprout. Expect sprouting in 3 to 5 days with a preferred soil temperature of 75° to 85°F. Transplant to the garden when the plant is 3 to 5 inches tall.

Growing Tips

Zinnias are happiest in hot, blasting, all-day sun. When grown in shady conditions, they are predisposed to mildew, growing leggy and producing fewer blooms. Flower-support netting is recommended for the top-heavy canopy of blooms that can easily fall over in the garden. I stop all fertilization once the foliage grows through the support nettings. I have found that fertilizing in the second half of zinnias' life cycle seems to fuel mildew more than it does the plants. The organic dry fertilizer added when the bed is prepared will carry the plants through the rest of the season.

'Benary's Giant', from top left to right: scarlet, deep red, coral, orange, white, salmon rose, golden yellow, wine, carmine rose, lilac, purple, bright pink.

ABOVE: *This is a small but powerful flower! 'Oklahoma' zinnias produce 2-inch blooms that are fully doubles. They are my favorites.*

RIGHT: *Rely on 'Benary's Lime Green' to have a double row of petals. The lime-green color makes it useful in any bouquet or garden.*

OPPOSITE: *The Queen series produces 2- to 3-inch blooms with a high percentage of double blooms. Their antique coloring has made them a popular cut flower.*

Harvesting Tips

Cut when all the flower petals are open and as a few of the star-like yellow stamens in the center of the flower are beginning to open. Flowers will not open or develop any further after the stem is cut. Harvest too early and the stems will droop. Make the first cut of the central stem almost at ground level, just above three to four side shoots. This sets up the plant to produce a season of strong, tall stems. Make the subsequent harvest cuts at the end of the stem being harvested. Remove all foliage except for the top two leaves. Blooms last 7 to 10 days in a vase.

Favorite Varieties

The Benary's Giant zinnias are the most widely grown zinnia. They produce the most voluptuous 4- to 5-inch blooms, are prolific, and are mildew resistant. They're available in 13 colors, including lime green. Other favorites include 'Uproar Rose', 'Queen Red Lime', 'Queen Lime Green', 'Oklahoma', 'Giant Cactus', 'Zowie!', and 'Cupcake'.

SUNFLOWERS, *Helianthus annuus*

AT A GLANCE

Sowing seed: Start indoors or in the garden

Sun: 6 or more hours

Height: 36" to 72"

Rows per 36" bed: Depends on size of blooms desired; see Growing Tips below

Spacing in the row: 6" to 24"

Days to bloom: 55 to 110

Deer resistance: Moderate

Container use: Excellent with shorter varieties

Visitors to expect: Butterflies, bees, and other pollinators, songbirds, and beneficial insects

The ProCut series orange is the sunflower I grow most often. It is the traditional color with a strong neck and a long vase life.

The 'Teddy Bear' is a dwarf growing to 36 inches in the garden and 24 inches in a container, making it a great addition to the garden. PHOTO BY GARDENER'S WORKSHOP FARM

Sometimes called the chocolate sunflower, 'Moulin Rouge' is stunning. This is a branching sunflower that produces black-oil sunflower seeds—a favorite of birds.

LISA'S GROUND RULES

Seed-Starting Tips

Sunflower seeds can be started indoors for an early start or planted out in the garden. For soil blocking, use the 2-inch blocker, pushing the seed into the block ½ inch. For starting in a small container or in the garden, cover the seed with ½ inch of soil. Sunflower seeds need darkness to sprout. Expect sprouting in 3 to 5 days with a preferred soil temperature of 75° to 85°F. Sunflowers grow quickly and resent becoming rootbound in a container. We use plug trays to start, and the seedlings are transplanted to the garden when they reach 3 to 5 inches tall, which is 2 to 3 weeks from the seed sowing time.

Growing Tips

Sunflowers are strong and easy growers, and I have found their threshold for the minimum amount of care required to still grow great flowers. We plant transplants into prepared beds that have had dry organic fertilizer applied per package directions. The transplants are hand-watered at planting and two more times over the next 10 days if there is no rain. Even in our long, hot, humid summers, these plants grow with literally no intervention from us. Flower-support netting is recommended for the top-heavy canopy of blooms that can easily fall over in the garden. Rabbits and birds are

'Starburst Lemon Aura' is a branching sunflower that produces lemon-yellow double blooms. It makes an excellent cut flower or, if left in the garden, the birds enjoy its seed. PHOTO BY GARDENER'S WORKSHOP FARM

ProCut series lemon has a lighter-brown center, and its petals are the clearest lemon yellow. They brighten a bouquet and look terrific in the garden.

ProCut series bicolor is a great addition to the late summer and fall garden. The band of rusty brown on the petals says fall!

especially fond of young sunflower transplants; using a floating row cover will eliminate this threat. See Chapter 8 for more on using row covers. **Single-stemmed varieties'** bloom size is determined by the distance given between the plants in the garden. For a bouquet-size sunflower (4 to 5 inches), plant four rows in a 36-inch bed with 6 inches between the plants in the row. The more space between the single-stem variety, the larger the bloom will grow. **Branching sunflowers** require more garden space because they grow many branches over time. Placing them 12 to 24 inches apart in the garden will allow for maximum branching. Pinch the central stem back to four or five side shoots when the plant reaches 24 inches.

Harvesting Tips

Harvest sunflowers as the blooms are beginning to open. Flowers will continue to open and develop after the stem is harvested. This timely harvest will prevent grasshoppers, beetles, and other insects from damaging the petals. Single-stemmed varieties may be cut at any length. On branching varieties, make the cut at the base of the stem being cut. Remove all foliage below the top leaf just below the bloom. Single-stemmed blooms last 10 to 14 days, and branching blooms last 5 to 7 days in a vase. The novelty-colored branching varieties are known to drop their petals prematurely—harvesting when the first petal lifts off of the center disk of the flower can help prevent this. Many of the sunflower varieties are pollenless, which makes them a longer-lasting cut flower. Pollenless blooms do produce nectar.

Favorite Varieties

Single-stemmed pollenless favorites are the ProCut and Sunrich series, which are each available in several colors. Branching pollenless favorites include 'Moulin Rouge' and 'Starburst Lemon Aura'. 'Teddy Bear' is a branching dwarf that does great in containers and small gardens while producing minimal pollen. Not intended as a cut flower but for a giant garden sunflower, 'Mammoth' is a single-stemmed variety that produces pollen. See page 44 for giant-sunflower growing tips.

CELOSIA (COCKSCOMB AND PLUMES),
Celosia argentea var. *cristata* and
C. argentea var. *plumosa*

AT A GLANCE

Sowing seed: Start indoors or in the garden

Sun: 6 or more hours

Height: 24" to 48"

Rows per 36" bed: 3 to 4

Spacing in the row: 6" to 12"

Days to bloom: 90 to 120

Deer resistance: Excellent

Container use: Excellent with shorter varieties

Visitors to expect: Butterflies, bees, and other pollinators, songbirds, and beneficial insects

'Pampas Plume' comes in a mix of beautiful colors, perfect in a small garden. Keeping the patch well harvested will help deter developing seed.

'Chief Mix' is a top producer on our farm because it gives loads of blooms per plant. The mix includes orange, carmine, gold, fire, and red-orange.

LISA'S GROUND RULES

Seed-Starting Tip

I find it more successful to start the tiny seeds of celosia indoors. For soil blocking, use the ¾-inch blocker or start in a small container. Firmly seat the seed on the surface of the soil. Do not cover the seed with soil. Celosia seeds need light to sprout. Expect sprouting in 5 to 10 days with a preferred soil temperature of 75° to 85°F. Transplant to the garden when the plant is 3 to 5 inches tall.

Growing Tips

This is a huge family of plants that offers various bloom shapes, colors, and heights. Because they tend to become stunted when the roots are potbound, they do best started from seed versus purchasing transplants. Celosia does exceptionally well started in soil blocks because it doesn't become rootbound. Flower-support netting is recommended for the top-heavy canopy of blooms that can easily fall over in the garden. Cockscomb is a favorite overnight hangout for bumblebees.

Harvesting Tips

Harvest blooms when they are fully developed and before they begin to make seed. Flowers do not develop any further once the stem is harvested. Make the first cut of the central stem almost at ground level, just above three to four side shoots. This sets up the plant to produce more strong, tall stems. Leave the foliage on the top third of the stem. Typically celosia foliage is very attractive and does not interfere with hydrating the stem. Make the subsequent harvest cuts at the end of the stem being harvested. Blooms last 10 to 14 days in a vase.

Favorite Varieties

Favorite plumes are the Sunday, Castle, and Century series, 'Sylphid', and 'Hi-Z'. Cockscomb varieties include the Chief and Cramer's series, 'Spring Green', and 'Jura Salmon'. If blooms are left on the plant to develop seed, they will reseed themselves for next season.

'Spring Green' cockscomb produces one large bloom per plant. Do not pinch the transplant. Note: Plant 6 inches apart in all directions, closer than branching types.

The Sunday series produces thick, beautiful plumes in gorgeous colors. The plants produce many stems per plant throughout the season. Orange is my favorite!

GIANT MARIGOLDS, *Tagetes erecta*

AT A GLANCE

Sowing seed: Start indoors or in the garden

Sun: 6 or more hours

Height: 36"

Rows per 36" bed: 3 to 4

Spacing in the row: 6" to 12"

Days to bloom: 70 to 90

Deer resistance: Excellent

Container use: Moderate

Visitors to expect: Songbirds, wasps, and other beneficial insects

LISA'S GROUND RULES

Seed-Starting Tips

Marigold seeds can be started indoors for an early start or planted out in the garden. For soil blocking, use the ¾-inch blocker, sticking the pointy end of the seed into the block until just the tail of the seed is sticking out. For starting in a small container or in the garden, cover the seed with ¼ inch of soil. Marigold seeds need darkness to sprout. Expect sprouting in 3 to 5 days with a preferred soil temperature of 75° to 85°F. Transplant to the garden when the plant is 3 to 5 inches tall.

Growing Tips

Marigolds are not fussy about their growing conditions. Give them sunshine and some space and they will go all summer. To prevent the central stem from growing into a huge, multibloom stem that is not very useable as a cut flower, pinch it out above the bottom three to four side shoots. This will set up the plant to grow an abundance of perfect-size side shoots. In addition to being a nectar and pollen source, marigolds usually draw many beneficial bugs. The bugs seem to love the ribbon blooms for sleeping and watching. Marigolds are also an excellent trap crop for Japanese beetles (see Chapter 8 for more). Flower-support netting is recommended for these tall stems.

Harvesting Tips

Harvest marigold stems any time after the blooms are halfway open. When left in the garden to mature and develop further, they become more beautiful, but the risk of insect and weather damage to petals increases.

The Giant marigold series is true to the name; they are huge and productive, a late-season crop that is perfect for fall.

Flowers will continue to open and develop after the stem is harvested. Make the first cut of the central stem (if not pinched out earlier) almost at ground level, just above three to four side shoots. This sets up the plant to produce more strong, tall stems. Remove all foliage below the top set of leaves below the bloom. Make the subsequent harvest cuts at the end of the stem being harvested. Blooms last 10 to 20 days in a vase. It is not uncommon for stems to root in the vase.

Favorite Varieties

The most prolific bloomers with strong stems are the Giant and Jedi series and 'Babuda'. Marigolds will bloom right up to frost, and the yellow, gold, and orange color selections fit right in with the fall season.

BASIL, *Ocimum basilicum*

AT A GLANCE

Sowing seed: Start indoors or in the garden

Sun: 6 or more hours

Height: 24" to 30"

Rows per 36" bed: 3 to 4

Spacing in the row: 6" to 12"

Days to bloom: 60 to 65

Deer resistance: Excellent

Container use: Excellent

Visitors to expect: Butterflies, bees, and other pollinators, songbirds, and beneficial insects

LISA'S GROUND RULES

Seed-Starting Tips

Seeds are easily started indoors or planted directly in the garden. For the first planting in spring, start seeds indoors for the earliest start, and sow later plantings directly in the garden. For soil blocking, use the ¾-inch blocker, pushing the seed ⅛ inch deep into the block. For planting in a small container or in the garden, cover the seed with ⅛ inch of soil. Basil seeds will sprout with or without light. Expect sprouting in 5 to 10 days with a preferred soil temperature of 75° to 85°F. Transplant to the garden when the plant is 3 to 5 inches tall.

'Mrs. Burns' Lemon' basil on the left and 'Cinnamon' basil on the right. This fragrant foliage really pushes a beautiful bouquet over the top.

Growing Tips

I find planting only the number of plants that I can keep harvested weekly leads to a longer harvest period and better quality cut-flower stems. Cutting the stems as they mature prevents them from becoming woody and allows the stem to drink water more freely. However, the benefit of having a few plants left in the garden to bloom is that the bees love basil blooms. Flower-support netting is recommended for these tall stems, which get top heavy with their canopy of foliage.

Harvesting Tips

Basil is used as a fragrant foliage in bouquets. It is best to harvest early in the morning before the heat of the day sets in. Harvest when the first few flowers at the bottom of the bloom begin to open. Make the first cut of the central stem almost at ground level, just above three to four side shoots. This sets up the plant to produce more strong, tall stems. Remove all foliage that will fall below water level. Make the subsequent harvest cuts at the end of the stem being harvested. Foliage lasts 7 to 10 days in a vase. If the foliage is still wilted the morning after harvest, remove more leaves and/or harvest earlier in the morning next time. Harvesting basil when it is still too immature will result in wilting that cannot be restored.

Favorite Varieties

My hands-down favorite basil for a bouquet is 'Mrs. Burns' Lemon'. It smells just like fresh lemons, and the foliage is a beautiful light green that brightens bouquets. 'Cinnamon' basil is a second runner-up, with a darker red stem and pinkish bloom.

MEXICAN SUNFLOWERS, *Tithonia* spp.

AT A GLANCE

Sowing seed: Start indoors or in the garden

Sun: 6 or more hours

Height: 36" to 72"

Rows per 36" bed: 2

Spacing in the row: 12" to 24"

Days to bloom: 100 to 120

Deer resistance: Excellent

Container use: Excellent with shorter variety

Visitors to expect: Butterflies, bees, and other pollinators and beneficial insects

LISA'S GROUND RULES

Seed-Starting Tips

It is best to start seeds indoors because of the long growing period required before blooming. For soil blocking, use the ¾-inch blocker, sticking the pointy end of the seed into the block until just the tail of the seed is sticking out. For planting in a small container or in the garden, cover the seed with ⅛ inch of soil. Mexican sunflowers need darkness to sprout. Expect sprouting in 5 to 10 days with a preferred soil temperature of 75° to 85°F. Transplant to the garden when the plant is 3 to 5 inches tall.

Growing Tips

These late-summer bloomers are a favorite in bouquets and of the monarch butterfly. Space these plants 24 inches apart when grown for an ornamental and 12 inches apart when harvesting regularly for cut flowers. Flower-support netting is recommended for these tall plants that quickly get top heavy. To maximize branching and also help keep the plant a little shorter, pinch the central stem back to just above the lowest four or five side shoots when the plant reaches 24 inches.

Harvesting Tips

Harvest as the blooms are just beginning to open. Flowers will continue to open and develop after the stem is harvested. This timely harvest will prevent grasshoppers, beetles, and other insects from damaging the petals. Make the first cut of the central stem (if not pinched out earlier) almost at ground level, just above three to four side shoots. This sets up the plant to produce a season of strong, tall stems. Make the subsequent harvest cuts at the end of the stem being harvested. Blooms last 5 to 7 days. Remove all foliage below the top set of leaves just below the bloom. If it is still wilted the following morning, remove more foliage and/or harvest earlier in the day next time.

Favorite Varieties

'Torch' is the favorite variety for cutting, while 'Goldfinger' is a dwarf that can be used in containers and also makes a suitable cut flower.

Mexican sunflower 'Torch' is over 6 feet tall in our garden and is covered in blooms and pollinators. Here is a native bee going for the gold!

FLOSS FLOWERS, *Ageratum houstonianum*

AT A GLANCE

Sowing seed: Start indoors

Sun: 6 or more hours

Height: 24″ to 36″

Rows per 36″ bed: 3 to 4

Spacing in the row: 6″ to 12″

Days to bloom: 80 to 100

Deer resistance: Excellent

Container use: Excellent

Visitors to expect: Butterflies, bees, and other pollinators

LISA'S GROUND RULES

Seed-Starting Tips

Floss flower prefers to be started indoors. For soil blocking, use the ¾-inch blocker or start in a small container. Firmly seat the seed on the surface of the soil and do not cover the seed with soil. Floss flower needs light to sprout. Expect sprouting in 5 to 10 days with a preferred soil temperature of 75° to 85°F. Transplant to the garden when the plant is 3 to 5 inches tall.

Growing Tips

Blue flowers bring a bouquet to life like no other flower, and floss flower is easy to grow. Some varieties are frequently available at local nurseries, but be aware that most of those are the short-bedding plants. A second planting is easily rooted from a 3- to 4-inch piece of stem. Remove all but one or two leaves, dip the stem end into rooting-hormone powder, and place it in moist potting soil. Keep warm and moist and it will root faster than seeds can sprout and grow. Plant in the garden once roots develop. Flower-support netting is recommended for these tall plants that quickly get top heavy with foliage and blooms.

Harvesting Tips

Cut when the flower buds are blue but not fuzzy yet. Flowers will continue to develop further after the stem is harvested. Make the first cut of the central stem almost at ground level, just above three to four side shoots. This sets up the plant to produce a season of strong, tall stems. Make the subsequent harvest cuts at the end of the stem being harvested. Remove all foliage except for the top two to four leaves. Blooms last 7 to 10 days in a

Blue flowers bring brilliance to a bouquet and garden like no other. Ageratum is easy to keep blooming all summer and fall when harvested weekly.

vase. The deeper and more frequently you harvest this plant, the more it will bloom; it's truly a cut-and-come-again-and-again plant.

Favorite Varieties

'Blue Horizon' and 'Tall Blue Planet' are the favorites in the cut-flower gardening circles. Floss flowers are also available in white and red blooms, but I have found both to be fairly puny garden plants and cut flowers.

GRASSES, *Setaria italica*, *Pennisetum glaucum*, *Panicum elegans*, *Triticum durum*, and *Triticosecale* spp.

AT A GLANCE

Sowing seed: Start indoors or in the garden

Sun: 6 or more hours

Height: 28" to 60"

Rows per 36" bed: 4 to 6

Spacing in the row: 6"

Days to bloom: 60 to 80

Deer resistance: Poor

Container use: Excellent

Visitors to expect: Songbirds and beneficial insects

LISA'S GROUND RULES

Seed-Starting Tips

Grass seeds can be started indoors for an early start or planted out in the garden. For soil blocking, use the 2-inch blocker and push the seed ¼ inch deep into the block. In a small container or in the garden, cover the seed with ¼ inch of soil. Grass seed needs darkness to sprout. Expect sprouting in 3 to 5 days with a preferred soil temperature of 75° to 85°F. Transplant to the garden when the plant is 3 to 5 inches tall.

Growing Tips

When growing the more costly hybrid seeds, such as 'Purple Majesty' (*Pennisetum glaucum*), I always start the seed indoors to conserve seeds. For some of the other types of grasses, such as 'Black Tip Wheat' (*Triticum durum*) or 'Lime Light Spray' millet (*Setaria italica*), that aren't so costly, sowing in the garden works well. No flower-support netting is needed.

Harvesting Tips

Grasses can be harvested as the head is just emerging for fresh use. They can also be left in the garden to move through the stages of growing a larger head and shedding pollen to beginning to dry. 'Purple Majesty', 'Lime Light Spray' millet, and 'Frosted Explosion' (*Panicum elegans*) are all best harvested just as the top 3 to 4 inches of the head emerges. Make the cut at ground level to encourage more stems to develop. For fresh use of bunching grasses, such as wheat and rye, I grab many stems in my hand and use a sod-cutter knife to cut the

'Lime Light Spray' millet is a great addition to bouquets. Its lime-green color is so versatile that we plant it weekly alongside our sunflower successions.

stems. Holding the stems near the heads, I shake off any loose foliage and use my fingers to comb out any unwanted leaf matter. For dried use, allow them to continue in the garden until they start to dry.

Favorite Varieties

I tend to favor 'Purple Majesty' and 'Lime Light Spray' millets because they give multiple stems per plant over several weeks. Some others that tend to be a one-time harvest are 'Black Wheat', 'Silver Tip Wheat', and 'Frosted Explosion'.

'Purple Majesty' millet resembles corn-stalk foliage with a cattail head all in maroon. A very productive plant with regular harvests, it is a bird favorite.

A garden patch of cosmos is beautiful. Do not feed cosmos at any stage of life. Fertilization leads to weak foliage and few flowers.

CHAPTER 4

COSMOS, *Cosmos bipinnatus*

AT A GLANCE

Sowing seed: Start indoors or in the garden

Sun: 6 or more hours

Height: 36" to 48"

Rows per 36" bed: 3 to 4

Spacing in the row: 6" to 12"

Days to bloom: 60 to 90

Deer resistance: Excellent

Container use: Excellent with shorter varieties

Visitors to expect: Butterflies, bees, and other pollinators, beneficial insects, and songbirds

LISA'S GROUND RULES

Seed-Starting Tips

Cosmos seeds can be started indoors for an early start or planted out in the garden. For soil blocking, use the ¾-inch blocker, sticking the pointy end of the seed into the block until just the tail of the seed is sticking out. For planting in the garden or in a small container, cover the seed with ¼ inch of soil. Cosmos seed needs darkness to sprout. Expect sprouting in 5 to 10 days with a preferred soil temperature of 75° to 85°F. Transplant to the garden when the plant is 3 to 5 inches tall.

Growing Tips

Cosmos will bloom all summer, but they really love the conditions of late summer and fall—be sure to plant some in mid- to late summer. They prefer to grow in average soil, meaning when the soil is rich in compost and organic matter, it tends to make cosmos foliage weak and leggy. Flower-support netting is recommended for these tall plants that can quickly and easily go down in the garden.

Harvesting Tips

Harvest as the blooms are just cracking open. Flowers will continue to open and develop after the stem is harvested. This timely harvest will prevent beetles and other insects from damaging the delicate petals. Make the first cut of the central stem almost at ground level, just above three to four side shoots. This sets up the plant to produce a season of strong, tall stems. Make the subsequent harvest cuts at the end of the stem being harvested. Blooms last 5 to 7 days. Remove all foliage

A mix of cosmos 'Double Click', 'Sonata', and 'Cupcake'. Be sure to harvest the blooms early as they have a short vase life.

below the top set of leaves just below the bloom. If it is still wilted the following morning, remove more foliage and/or harvest earlier in the day next time.

Favorite Varieties

Cosmos have several different bloom styles, with most having similar colors. The Double Click series has rows and rows of petals, 'Sonata' has fluted petals, and 'Cupcake' has the look of a poppy.

Gomphrena QIS is an excellent garden plant and cut flower. It is drought tolerant and the blooms are long lasting. A pollinator favorite.

GLOBE AMARANTH,
Gomphrena globose and *G. haageana*

AT A GLANCE

Sowing seed: Start indoors

Sun: 6 or more hours

Height: 18" to 30"

Rows per 36" bed: 3

Spacing in the row: 6" to 12"

Days to bloom: 85 to 100

Deer resistance: Excellent

Container use: Excellent

Visitors to expect: Butterflies, bees, and other pollinators and beneficial insects

LISA'S GROUND RULES
Seed-Starting Tips
Seeds should be started indoors. For soil blocking, use the ¾-inch blocker or start in a small container. Firmly seat the seed on the surface of the soil. Do not cover the seed with soil. Globe amaranth needs light to sprout. Expect sprouting in 5 to 10 days with a preferred soil temperature of 75° to 85°F. Transplant to the garden when the plant is 3 to 5 inches tall.

Growing Tips
This showy garden plant produces an abundance of little round blooms that make excellent fresh and dried cut flowers. Butterflies and small pollinators are particularly drawn to this heat-tolerant plant. No flower-support netting is recommended.

Harvesting Tips
Harvest when the clover-like blooms are fully developed. Flowers will not develop any further once the stem is harvested. Cut an entire branching stem by going all the way to ground level to make the cut. This sets up the plant to produce a season of stems. Make the subsequent harvest cuts at the base of the branching stem being harvested. Blooms will last 10 to 14 days in water and seem to last forever when air-dried. Remove foliage that will fall below the water level. For air-drying, remove foliage from the bottom two-thirds of the stems, bunch them, and rubber band together. Hang to dry in a cool, dark area. The blooms hold their color indefinitely.

Gomphrena is an everlasting flower, which means it can be air dried and holds its color. Pictured here is gomphrena 'Carmine', used on a party collar for Beri.

Favorite Varieties
There are several cultivars of globe amaranth. My favorites are the QIS series (*G. haageana*), which includes 'Strawberry Fields' and 'Carmine', coming in at 30 inches tall with straighter single stems that are easier to harvest. The Tall series (*G. globosa*) has a bold purple that has many branches on a stem, which is more of a challenge, but the color is dynamite!

Cool-Season Hardy Annuals

Gardeners everywhere can reap the beauty and benefits of cool-season hardy annuals. Hardy annuals bring flowers on earlier, and with these early flowers, beneficial insects are invited to come and set up housekeeping for the season. Even gardens in areas with long, hot, humid summers (like mine) and in northern areas with long, frigid winters can grow hardy annuals; it's just that we plant them at different times. Success with cool-season hardy annuals is all about timing. Finding the hardy-annual-planting sweet spot for your garden will lead to years of abundant early rewards.

Many gardeners forgo planting hardy annuals because they seem fussy and difficult to please, but nothing could be further from the truth. The key to success is planting them at their preferred time, which is when it is cool to even cold—so different from what we would expect. Hardy annuals, sometimes called winter annuals, can become some of the easiest plants to grow once you give them what they want.

The misfortune this family of plants fell upon is that plants are rarely available for purchase at their optimal planting time. Most often they are found on racks for the spring shopping season. This is tragic because, for most gardens, mid-spring is way too late to plant them. Planting hardy annuals late means the plants are barely established before they must start growing foliage and flowers. Soon after, summer-like weather arrives with higher temperatures and humidity, and the plants begin to struggle. Their decline is swift because they don't have an established root system to support them. This sets cool-season hardy annual plants up for a troublesome start and an early demise. But it doesn't have to go this way.

OPPOSITE: *Snapdragon 'Chantilly', bachelor buttons 'Boy', dill 'Bouquet', pot marigold 'Pacific Beauty', poppies 'Champagne Bubbles' and 'Giant Pod', pincushion flower 'Black Knight', bupleurum, and orlaya.*

The pot marigold 'Pacific Beauty' thrives in spring's cool conditions. One of the earliest flowers to bloom on my farm and a favorite of pollinators.

WHEN SHOULD I PLANT HARDY ANNUALS?

In regions that have long and cold winters with a heavy snow load, hardy annuals can be planted in the garden in *early* spring. This is weeks earlier than their well-known cousins, warm-season annuals, can be planted. Those gardening in milder winter regions, such as my Zone 7, can plant in the fall. The young plants will winter over to become some of the earliest spring bloomers. If space allows, you can plant again in early spring to extend the blooming season.

To learn when to plant hardy annuals in your garden, find your

- **winter hardiness zone** (consult the USDA website at planthardiness.ars.usda.gov)

- **first and last expected frost dates** (contact your local cooperative extension service or master gardener organization)

Any hardy annual that *is not* winter hardy in your zone should be planted in early spring because it will not survive winter. For example, a gardener in Zone 4 would plant sweet peas in early spring because they are only hardy to Zone 7.

Any hardy annual that *is* winter hardy in your zone or one with a lower zone number than yours can survive winter and be planted in the fall. A garden in Zone 6, for example, can plant rudbeckia in the fall to winter over because it is hardy to Zone 5.

When is early spring planting? Plant 6 to 8 weeks *before* the last expected spring frost.

When is fall planting? Plant 6 to 8 weeks before the first expected fall frost.

WHY IS PLANTING ON TIME SO IMPORTANT?

When hardy annuals are planted *into their preferred conditions*—cool to cold soil and air temperatures—they thrive. When plants thrive, they naturally grow robust roots. The roots of a plant are the foundation that delivers all the nutrients, moisture, and everything else it needs. They are a plant's lifeline, really. With a strong foundation, the plant can face and overcome adverse conditions like heat and humidity and better resist disease and pests. This means that the plant can produce longer into the warm summer season.

When hardy annuals are planted *out of their preferred conditions*, there isn't enough a gardener can do to overcome the adverse conditions of the season. Every year I miss the window of opportunity with at least one planting of hardy annuals, and it shows like a red flag in my garden. The plants limp along, are not robust, and are slower to produce. They produce shorter and fewer stems. There is no comparison of the quality of a hardy annual planted into its preferred element to the quality of the same variety of plant planted out of it.

A common question I receive from gardeners once spring is in full swing is "I didn't get to plant my hardy annual seeds yet. Is it okay to start now?" My advice is that you save the seeds until their next planting season—either in the fall or early spring. I recommend paperclipping the seed packets to your calendar on the month as a reminder. If you plant too late, not only will the plants not thrive, but the experience will become a discouragement that will likely squash your desire to grow hardy annuals again in the future.

Leaf lettuce that was fall-planted and harvested multiple times held on through winter under row cover and regrew in spring to offer more harvests.

HARDY-ANNUAL PLANTING NOTES

Because cool-season plants are planted into cold soil and cool weather they have additional requirements to thrive.

- Plant on time to give plants the time they need to get established.

- Plant into well-drained soil. A common cause for loss of hardy annuals is that the roots get waterlogged. Because the soil won't dry out as quickly in the cold months, excellent drainage is a must. Rain and snow can keep a poor drainage site soggy all winter and early spring.

- Raised beds are beneficial because the soil warms up sooner and drainage is improved.

EARLY-SPRING PLANTING NOTES

Early-spring and fall planting times each offer their own challenges and benefits. For spring:

- Prepare the area in fall for the early-spring planting. Soil is usually too wet or snow covered to be prepared in early spring, and that's a common cause of planting delays. Prepare and cover or mulch the area to wait for the planting in early spring. Covering or mulching will deter cool-season weeds from sprouting. See Chapter 8 for more on deterring weeds.

- Always choose to plant transplants when possible in early spring versus sowing seeds in the garden. It is more difficult to get seeds to sprout outdoors due to cold conditions and possible snow. For those seeds that prefer being planted and sprouted in the garden, use a row cover to aid in warming the soil to increase seed sprouting. See Chapter 8 for more on using row covers.

- Using hoops and row covers in early spring will go a long way in protecting young transplants from spring winds and hungry bunnies.

FALL PLANTING NOTES

- Always choose fall planting if a plant is hardy in your winter hardiness zone.

- The 6- to 8-week planting window is for planting seeds directly in the garden and planting

Lightweight row cover used with hoops and weight bags protects young plants from rabbits, deer, wind, cold, insect pressure, and other environmental issues.

transplants. This timeframe allows seeds to sprout and grow to the desired 4- to 8-inch-size seedling or for the transplant to become established before winter-like weather begins.

- These young plants spend the fall, winter, and early spring growing a strong root system that will carry them through the following season like a champ. Do not expect foliage growth during winter. In fact, they will look like little frozen popsicles for a good part of it.

- Using hoops and row covers on fall plantings all winter will protect young transplants from winter winds and grazing deer.

SWEET PEAS, *Lathyrus odoratus*

AT A GLANCE

Sowing seed: Start indoors or in the garden

Winter hardiness zone: 7

Sun: 6 or more hours

Height: 36" to 72" vine

Rows per 36" bed: 1

Spacing in the row: 6" to 12"

Days to bloom: 75 to 85

Deer resistance: Excellent

Container use: Excellent with shorter varieties

Visitors to expect: Butterflies, bees, and other pollinators and beneficial insects

LISA'S GROUND RULES

Seed-Starting Tips

I find the greatest success starting sweet-pea seeds indoors. Soak the seeds for 1 to 3 hours before planting to soften the outer shell of the seed for quicker sprouting. For soil blocking, use the 2-inch blocker. For starting in a container, use a 3-inch pot. Cover the seed with ¼ inch of soil. Expect sprouting in 7 to 10 days with a preferred soil temperature of 70° to 75°F. Sweet peas resent becoming potbound. Transplant to the garden as soon as possible when the vine is 4 to 6 inches tall.

Growing Tips

Planting in garden soil that is rich in organic matter goes a long way in helping retain moisture and deliver nutrition that sweet peas thrive on. Vines grow best on a trellis that the tendrils can wrap around as they climb. Help the vines to get started by wrapping them around the trellis as they grow.

This vase filled with pink sweet peas from the 'Romance Mix' was a gift to me from Steve's mom. She received it in 1947 as a wedding gift.

Harvesting Tips

The fragrant blossoms of sweet peas should be harvested at least every other day to keep the vines producing. Blooms left on the vine will develop into seed pods, and blooming will cease. Harvest when at least half of the blooms are open. Flowers do not continue to open once the stem is cut. For individual stems, cut at the base of the stem. For the showiest and tallest stems, cut a vine that has several stems blooming—it will regrow. Remove all foliage that will fall below the water level in the vase. Blooms last 5 to 7 days in a vase. Seeds, pods, and plant are poisonous.

Favorite Varieties

'Royal Family Mix' has been the standard in our garden for years. It includes a wide variety of colors, has a nice fragrance, and is a strong grower. In containers, 'Knee-Hi Mix' works well, with a 36-inch vine that drapes nicely. Becoming more widely available are English sweet peas, which have larger, ruffled blooms and longer stems. Some favorites include 'Romance Mix' (pastel mix), 'Just Julia' (blue), 'Mumsie' (hot pink), and 'Castlewellan' (peach).

SNAPDRAGONS, *Antirrhinum majus*

The Rocket series is one of the latest varieties to begin blooming and blooms into our hot summer. Colors include yellow, pink, red, and white.

LISA'S GROUND RULES

Seed-Starting Tips

Snapdragon seeds should be started indoors. For soil blocking, use the ¾-inch blocker. For starting in a container, use a small-cell container. Firmly seat the seed on the surface of the soil and do not cover the seed with soil—it needs light to sprout. Expect sprouting in 7 to 14 days with a preferred soil temperature of 70° to 75°F. Transplant to the garden when the plant is 3 to 5 inches tall.

Growing Tips

Snaps are one of the easiest and strongest hardy annuals to grow. Their tall, towering stems laden with blooms can easily fall over in the garden, so flower-support netting is essential for straight stems.

Harvesting Tips

Harvest snapdragons at least twice a week. Cut when the first few blooms on the bottom of the stem have opened. Flowers will continue to open and develop after the stem is harvested. Make the first cut of the central stem almost at ground level, just above three to four side shoots. This sets up the plant to produce a season of strong, tall stems. Make the subsequent harvest cuts at the end of the stem being harvested. Remove the bottom two-thirds of foliage. Snaps will continue to produce well into summer, with their stems becoming shorter as the temperatures rise. Blooms last 7 to 10 days in a vase.

The ruffled petal in the center of 'Madame Butterfly' blooms sets them apart from other snaps. Colors include bronze, cherry, ivory, pink, rose, and yellow.

Favorite Varieties

There are many varieties of snapdragons. Some have varying bloom shapes and color combinations. Pay attention to which ones bloom early in spring and later. By growing more than one variety, you can extend the blooming season naturally. The following varieties are listed in the order of the start of bloom time in my garden from early spring to late spring: 'Chantilly', 'Madame Butterfly', 'Opus', 'Rocket', 'Snappy Tongue'. For container growing, plant 'Sonnet' blooms mid-spring.

'Opus' snapdragons are early-spring bloomers and bring the bumblebees flocking to the garden. This series has great colors that include bicolors seldom seen.

POT MARIGOLDS, *Calendula officinalis*

AT A GLANCE

Sowing seed: Start indoors or in the garden

Winter hardiness zone: 7

Sun: 6 or more hours

Height: 18" to 30"

Rows per 36" bed: 3 to 4

Spacing in the row: 6" to 12"

Days to bloom: 50 to 60

Deer resistance: Excellent

Container use: Excellent

Visitors to expect: Butterflies, bees, and other pollinators and beneficial insects

LISA'S GROUND RULES

Seed-Starting Tips

I prefer to start these large, funny-looking seeds indoors. For soil blocking, use the 2-inch blocker. For starting in a container, use a small-cell container. Cover the seed with ¼ inch of soil. Expect sprouting in 7 to 10 days with a preferred soil temperature of 70° to 75°F. Transplant to the garden when the plant is 3 to 5 inches tall.

Growing Tips

Pot marigolds are one of the very earliest flowers to bloom in our garden in the spring and are rich producers of pollen and nectar. Flower-support netting is not necessary. Don't despair if you find aphids. Pot marigolds are often grown as a trap crop because aphids prefer these sticky sap plants over others in the garden. Just give ladybugs, hoverflies, and lacewings time to move in and clean the plants up.

Harvesting Tips

Harvest pot marigolds at least twice a week. Cut when the bloom is just beginning to unfurl the petals. Flowers will continue to open and develop after the stem is harvested. Make each of the cuts at the base of the stem as close to ground level as possible—I actually cut into the soil if I can to get more length. This sets up the plant to produce the tallest stems possible. Remove all foliage from the stem for the maximum vase life. They will continue to produce until the temperatures and humidity climb. Blooms last 7 days in a vase.

The bright blooms of 'Pacific Beauty' are so welcome in spring when soft pastels typically dominate! Pot marigolds are a strong pollinator plant.

Favorite Varieties

The challenge is stem length with these flowers. I grow the Prince series because it has proven to have the tallest stems, often topping near 30 inches, but its blooms aren't as large as some others. The 'Pacific Beauty' has the largest blooms in fantastic colors but often are under 24 inches.

BEE BALM, *Monarda hybrida*

AT A GLANCE

Sowing seed: Start indoors or in the garden

Winter hardiness zone: 5

Sun: 6 or more hours

Height: 24" to 36"

Rows per 36" bed: 3 to 4

Spacing in the row: 6" to 12"

Days to bloom: 70 to 80

Deer resistance: Excellent

Container use: Excellent

Visitors to expect: Butterflies, hummingbirds, bees, birds, and other pollinators and beneficial insects

LISA'S GROUND RULES

Seed-Starting Tips

When planting in the fall, planting the seeds directly in the garden works well. For planting in early spring, starting seeds indoors is preferred. For soil blocking, use the ¾-inch blocker. For starting in a container, use a small-cell container. Barely cover the seed with ⅛ inch of soil. Expect sprouting in 7 to 14 days with a preferred soil temperature of 70° to 75°F. Transplant to the garden when the plant is 3 to 5 inches tall.

Growing Tips

Monarda hybrida is a quick grower from seed to bloom. It is more drought tolerant and resistant to mildew than other bee balms. Flower-support netting is recommended for the top-heavy canopy of blooms that can easily fall over in the garden. For those who can plant 'Lambada' in the fall, you can extend the bloom season by planting again in early spring.

Harvesting Tips

Harvest bee-balm stems any time after the blooms begin to turn a little pinkish. If harvested when too immature, the stems will wilt and cannot be restored. Flowers will continue to open and develop after the stem is harvested. Make the first cut of the central stem almost at ground level, just above three to four side shoots. This sets up the plant to produce more strong, tall stems. Remove all foliage below the top set of leaves below the bloom. Make the subsequent harvest cuts at the end of the stem being harvested. Blooms last 7 days in a vase.

Favorite Varieties

'Lambada' is the preferred variety. If blooms are left on the plant to develop seed, they can reseed themselves for next season. There are other varieties available, most of which are perennials. Many can be aggressive growers in the garden.

Bee balm is an excellent cut flower and adored by bumblebees. I like to harvest just as the pinkish buds start to show for more of a green-silver appearance.

BLACK-EYED SUSANS, *Rudbeckia hirta* and *R. maxima*

AT A GLANCE

Sowing seed: Start indoors or in the garden

Winter hardiness zone: 5

Sun: 6 or more hours

Height: 24" to 48"

Rows per 36" bed: 3 to 4

Spacing in the row: 6" to 12"

Days to bloom: 100 to 120

Deer resistance: Excellent

Container use: Excellent with shorter varieties

Visitors to expect: Butterflies, bees, and other pollinators and beneficial insects

The green-eyed Susan 'Prairie Sun' is one of the toughest plants I have grown. It continues to produce as long as the plant is harvested.

LISA'S GROUND RULES

Seed-Starting Tips

I find the most success with starting rudbeckia seeds indoors. For soil blocking, use the ¾-inch blocker. For starting in a container, use a small-cell container. Firmly seat the seed on the surface of the soil and do not cover with soil; it needs light to sprout. Expect sprouting in 7 to 14 days with a preferred soil temperature of 70° to 75°F. Transplant to the garden when the foliage is 3 to 5 inches long.

Growing Tips

Black-eyed Susans are not only excellent as garden members and cut flowers but are also a favorite of native bees. When fall planting is possible, rudbeckia blooms much earlier the following season. Flower-support netting is recommended for these tall stems, which produce tons of blooms. Slugs can eat the young, tender foliage, but the plants usually quickly outgrow the damage.

Harvesting Tips

Harvest rudbeckia stems any time after the blooms are halfway open. When left in the garden to mature and develop further, they become more beautiful, but the risk of insect and weather damage on petals increases. Flowers will continue to open and develop after the stem is harvested. Make the first cut of the central stem almost at ground level, just above three to four side shoots. This sets up the plant to produce more strong, tall stems. Remove all foliage below the top set of leaves below the bloom. Make the subsequent harvest cuts at the end of the stem being harvested. Blooms last 7 to 14 days in a vase.

Favorite Varieties

There are many varieties of rudbeckia that vary in heights, bloom sizes, and color variations. My favorites are *R. hirta* 'Indian Summer' (36 to 48 inches), 'Double Daisy' (36 to 48 inches), 'Goldilocks' (24 to 36 inches), 'Prairie Sun' (24 to 36 inches) and *R. maxima* 'Giant' (36 to 60 inches). If blooms are left on the plant to develop seed, they can reseed themselves for next season.

LEFT: Rudbeckia maxima *looks like a sombrero and is huge! I love to cut it but also like to leave for the goldfinches.*

BELOW: *'Indian Summer' produces an abundance of blooms up to 6 inches across. It is an excellent producer that native bees are attracted to. Fall planting produces early blooms.*

POPPIES, *Papaver nudicaule* and *P. somniferum*

AT A GLANCE

Sowing seed: Plant seeds in the garden

Winter hardiness zone: 6

Sun: 6 or more hours

Height: 18" to 60"

Rows per 36" bed: 4

Spacing in the row: 6" to 9"

Days to bloom: 60

Deer resistance: Excellent

Container use: Excellent

Visitors to expect: Bees and other pollinators and beneficial insects

LISA'S GROUND RULES

Seed-Starting Tips

I find the most success with planting poppy seeds outdoors in the garden. The seeds are tiny, so don't do it on a windy day. Sowing them into a bed free of soil clumps will help germination. To plant, pour a few seeds into your palm and rub a few seeds between your fingers. Gently but firmly seat the seed into the soil, and water. Do not cover with soil. Expect sprouting in 7 to 14 days with a preferred soil temperature of 65° to 75°F.

Growing Tips

Monitor the young, sprouting, tender foliage for slug damage. See Chapter 8 for tips on slug control. No flower-support netting is needed.

Harvesting Tips

The Iceland poppy Champagne Bubbles series does not require special handling. Harvest the stems once the bloom is cracking open just enough to see what color the bloom will be. Flowers will continue to open and develop after the stem is harvested. Make each of the cuts at the base of the stem close to ground level—I actually cut into the soil if I can to get more length. This sets up the plant to produce tall stems. Remove all foliage. Blooms last 5 to 7 days in a vase. Harvest giant poppy pods soon after flower petals fall off. Make the cut near ground level to encourage more stems to grow. Remove all foliage. Pods can be used fresh or dried.

Giant poppy blooms are loved by native bees as well as honeybees. Leave the blooms in the garden until a large green pod has developed to harvest.

Favorite Varieties

The Champagne Bubbles series (*P. nudicaule*, 18 to 24 inches) produces stems tall enough for the vase and includes fabulous colors. The 'Giant Poppy Pod' (*P. somniferum*, 48 to 60 inches) is grown for the huge green pods that develop after the bloom fades; the blooms will not last as a cut flower. If pods are left on the plant to dry and burst, they can reseed.

'Champagne Bubbles' is a variety that needs no special handling to last in a vase. Harvest just as the buds begin to crack open.

DILL, *Anethum graveolens*

AT A GLANCE

Sowing seed: Plant seeds in the garden

Winter hardiness zone: 8

Sun: 6 or more hours

Height: 36" to 60"

Rows per 36" bed: 4

Spacing in the row: 6"

Days to bloom: 65 to 75

Deer resistance: Excellent

Container use: Excellent with shorter varieties

Visitors to expect: Butterflies, bees, and other pollinators and beneficial insects

LISA'S GROUND RULES

Seed-Starting Tips

I find the most success with planting dill seeds outdoors in the garden. Barely cover the seed with ⅛ inch of soil. Expect sprouting in 7 to 14 days with a preferred soil temperature of 60° to 70°F.

Growing Tips

While dill is a cool-season lover, it can do fairly well planted in early and late summer in an effort to extend the season. Once summer arrives, I plant in a garden that offers afternoon shade and keep it well watered. The stems won't be as tall as cool-season planting, but they will certainly be tall enough. The canopy of dill blooms will be full of beneficial insects and pollinators—I love visiting this patch just to watch. Flower-support netting is recommended for these tall stems, which produce tons of blooms.

Harvesting Tips

Harvest for the vase and for cooking. The foliage is especially beautiful in early-spring bouquets of poppies and bachelor buttons. Cut foliage at any stage. Harvest the blooms any time after half of the tiny yellow flowers are open. Make the cut just above the lowest three to four side shoots. This sets up the plant to produce more strong, tall stems. Remove all foliage below the top set of leaves below the bloom. Make the subsequent harvest cuts at the end of the stem being harvested. Blooms last 7 to 14 days in a vase. Cut-flower food will help the flower to continue opening. For blooms you intend to cook with, use plain water.

Blooms can be harvested as they turn yellow and up until they develop seeds. These well-established plants produce flowers well into summer.

Favorite Varieties

The best variety of dill for cut flowers is called 'Bouquet'. It has towering stems sometimes over 60 inches and produces an abundance of blooms. If blooms are left on the plant to develop seed, they can reseed themselves for next season.

PINCUSHION FLOWERS, *Scabiosa atropurpurea*

AT A GLANCE

Sowing seed: Start indoors or in the garden

Winter hardiness zone: 7

Sun: 6 or more hours

Height: 24" to 36"

Rows per 36" bed: 3 to 4

Spacing in the row: 6" to 12"

Days to bloom: 75 to 80

Deer resistance: Excellent

Container use: Excellent

Visitors to expect: Butterflies, hummingbirds, bees, and other pollinators and beneficial insects

LISA'S GROUND RULE

Seed-Starting Tips

I find the most success with starting pincushion seeds indoors. For soil blocking, use the ¾-inch or 2-inch blocker (some varieties have larger seeds than others). For starting in a container, use a small-cell container. Cover the seed with ¼ inch of soil. Expect sprouting in 7 to 14 days with a preferred soil temperature of 70° to 75°F. Transplant to the garden when the foliage is 3 to 5 inches long.

Pincushion flowers: 'Fire King', 'Black Knight', and 'Blue Cockade' are my favorites to grow. They make excellent cut flowers, and the bumblebees especially like them.

Growing Tips

Bumblebees love this flower! When fall planting is possible, they bloom much earlier in spring, inviting the bees into the garden. Planting in early spring can extend the bloom time well into summer, when tomato plants will benefit the most from the bumblebees' presence. Flower-support netting is recommended for these tall stems, which produce tons of blooms.

Harvesting Tips

Cut early in the morning before the bees come to work! Harvest stems when one-third to half of the tiny flowers are open. Flowers will continue to open and develop after the stem is harvested. Make the first cut of the central stem almost at ground level, just above three to four side shoots. This sets up the plant to produce more strong, tall stems. Remove all foliage below the top set of leaves below the bloom. Make the subsequent harvest cuts at the end of the stem being harvested. Blooms last 7 to 14 days in a vase. Cut-flower food will help the flower continue opening.

Favorite Varieties

There is a slight variation in blooming time, beginning with 'Black Knight', followed by 'Fire King' and 'Blue Cockade', with 'Salmon Queen' and 'Fata Morgana' starting last.

BACHELOR BUTTONS, *Centaurea cyanus*

AT A GLANCE

Sowing seed: Plant seeds in the garden

Winter hardiness zone: 6

Sun: 6 or more hours

Height: 24" to 36"

Rows per 36" bed: 3 to 4

Spacing in the row: 6" to 12"

Days to bloom: 65 to 75

Deer resistance: Excellent

Container use: Excellent

Visitors to expect: Butterflies, bees, and other pollinators and beneficial insects

LISA'S GROUND RULES

Seed-Starting Tips
Plant the seeds directly in the garden. Cover the seed with ¼ inch of soil. Expect sprouting in 7 to 14 days with a preferred soil temperature of 60° to 70°F.

Growing Tips
Bachelor buttons are easy and strong growers that can bloom in very early spring when planted in fall. They provide important habitat and nectar early in the season for native bees and others. Flower-support netting is recommended for these tall stems, which can easily go down in the garden.

Harvesting Tips
Harvesting will encourage blooming well into the heat of summer, long past their normal flash of blooms in spring! Cut the stem when the flower is just beginning to crack open. Allowing it to open indoors out of wind and sun increases the vividness of the colors. Flowers will continue to open and develop after the stem is harvested. Make the first cut of the central stem almost at ground level, just above three to four side shoots. This sets up the plant to produce more strong, tall stems. Remove all foliage below the top set of leaves below the bloom. Harvest early in the morning to help prevent wilting. If wilting persists, remove more foliage. Make the subsequent harvest cuts at the end of the stem being harvested. Blooms last 7 to 14 days in a vase. Cut-flower food will help the flower continue opening.

While 'Blue Boy' is the most well-known bachelor button, 'Pink Boy' always has a place in our garden. A very early bloomer in spring.

Favorite Varieties
The Boy series includes colors beyond the eye-catching 'Blue Boy', including 'Pinkie', 'Snowman', and a mix of the colors. If blooms are left on plant to develop seed, they can reseed themselves for next season.

FALSE QUEEN ANNE'S LACE,
Daucus carota and *Ammi majus*

AT A GLANCE

Sowing seed: Start indoors or in the garden

Winter hardiness zone: 6 (*Daucus carota*), 7 (*Ammi majus*)

Sun: 6 or more hours

Height: 36" to 48"

Rows per 36" bed: 3 to 4

Spacing in the row: 6" to 12"

Days to bloom: 75 to 110

Deer resistance: Excellent

Container use: Excellent with shorter varieties

Visitors to expect: Butterflies, bees, and other pollinators, beneficial insects, and songbirds

LISA'S GROUND RULES
Seed-Starting Tips

I find the most success with starting seeds indoors. For soil blocking, use the ¾-inch blocker. For starting in a container, use a small-cell container. For *Daucus carota*, cover the seed with ⅛ inch of soil; it needs darkness to sprout. For *Ammi majus*, firmly seat the seed on the surface and do not cover with soil; it needs light to sprout. For either type, expect sprouting in 7 to 14 days with a preferred soil temperature of 70° to 75°F. Transplant to the garden when the foliage is 3 to 5 inches long.

Growing Tips

This family of flowers makes excellent garden plants and cut flowers, but they are also the favorites of many beneficial insects. These flowers are a real favorite of hoverflies, whose offspring gobble up aphids. Flower-support netting is recommended for these tall stems, which can easily go down in the garden.

Harvesting Tips

Harvest the stems from when one-third of the tiny flowers are open up until they mature to setting seed. When left in the garden to mature and develop further, they become more beautiful, but the risk of insect and weather damage increases. Flowers will continue to open and develop after the stem is harvested. Make the first cut of the central stem almost at ground level, just above three to four side shoots. This sets up the plant to produce more strong, tall stems. Remove all foliage below the top

False Queen Anne's lace 'Dara' has a variation of colors. The blooms go through stages over time and all stages make great cut flowers.

set of leaves below the bloom. Make the subsequent harvest cuts at the end of the stem being harvested. Blooms last 7 to 14 days in a vase. Cut-flower food will help the flower to continue opening.

Favorite Varieties

For the large, flat, white flowers, select *A. majus* 'Graceland'. The pinkish blooms are *D. carota* 'Dara'. If blooms are left on the plant to develop seed, they can reseed themselves for next season.

BUPLEURUM, *Bupleurum rotundifolium* and *B. griffithii*

AT A GLANCE

Sowing seed: Plant seeds in the garden

Winter hardiness zone: 5

Sun: 6 or more hours

Height: 36" to 48"

Rows per 36" bed: 3 to 4

Spacing in the row: 6" to 12"

Days to bloom: 80 to 90

Deer resistance: Excellent

Container use: Good with shorter varieties

Visitors to expect: Bees and other pollinators and beneficial insects

LISA'S GROUND RULES

Seed-Starting Tips

Plant the seeds directly in the garden. Cover the seed with ¼ inch of soil. Expect sprouting in 10 to 20 days with a preferred soil temperature of 70° to 75°F. The newly emerged seedling is very thin and hard to see at first.

Growing Tips

Bupleurum (pronounced *boo-ploor-um*) is one of the toughest winter-hardy annuals. It smiles through the winter as a young plant up to Zone 5. It appreciates good drainage and moisture. It's an early bloomer that is very attractive to the smaller beneficial bugs and their larvae because of the cluster of tiny flowers. The thing is, tiny bugs have tiny mouths and need tiny flowers to eat from, a key element to inviting them into the garden early. Flower-support netting is recommended for these tall stems, which can easily go down in the garden.

Harvesting Tips

Harvest the stems from when one-third of the tiny flowers are open up until they mature to setting seed. The more flowers that are open, the longer the stem will last. In a vase, flowers will continue to open and develop after the stem is harvested. Make the first cut of the central stem just above the lowest 3 to 4 side shoots. This will encourage the side shoots to develop into longer, straighter stems. Remove all foliage except that on the top third of the stem. If wilting occurs, remove more

Bupleurum is a knockout cut flower that thrives in cool climates and will reseed in the garden. I never have enough of this flower to go around.

foliage. Make the subsequent harvest cuts at the end of the stem being harvested. Blooms last 10 to 14 days in a vase. Cut-flower food will help the flower to continue opening.

Bupleurum has clusters of tiny flowers that invite little native bees and wasps into the garden. Their tiny mouths require tiny flowers to feed from.

Favorite Varieties

B. rotundifolium 'Garibaldi' is the tallest and most robust variety. *B. griffithii* 'Green Gold' is also an excellent choice; it's a little shorter but still a great cut flower. If blooms are left on the plant to develop seed, they can reseed themselves for next season.

SECTION 3

Heroes of the Garden

Beneficial creatures are nature's way of pollinating plants, controlling pests, processing organic matter, dealing with weed seeds, and a whole lot more in a garden and yard. A smile is sure to spread across your face as you become familiar with some of these garden heroes and what tasks they are quietly doing without us even noticing. These guys are ready and waiting for the invite to come set up housekeeping, raise their families, and do their jobs. How do we invite them? By what we grow and how we treat it. Give them what they need and step back.

I consider a beneficial creature to be any living thing that contributes to my garden in some way. Every living creature benefits something at some time in their lifetime. What would the good bugs eat if there were no bad bugs? Did you know that possums eat ticks? The common starling, a bird people tend to loathe, eats slugs. I'm learning to not pass judgement on any creature until I know what they are, what their story is, and what they may contribute to my garden.

Pollinators and other beneficial insects, birds, reptiles, and other creatures are on the decline because of habitat loss, pollution, and pesticide poisoning. The yards and gardens we offer can invite them to stay, send them looking elsewhere, or perhaps even take their life—it's up to the gardener. Take a moment and view your garden and yard through the eyes of a bee, bird, or frog. What do you see? An all-natural garden of trees, shrubs, and plants to explore or an endless lawn with nowhere to hide, rest, eat, or raise a family?

Once we gain a better understanding of how the ecosystem works to support these creatures, it is easy to see how and why things have gone so badly for them. If you asked any beneficial insect or creature how life is going these days, I think they'd reply, "It seems like the world has declared war on us." One yard and garden at a time, this can change.

On the following pages, I hope you are going to reunite with some of the more well-known beneficial creatures and perhaps learn about some new ones. You might be skeptical or even frightened until you learn just what they are really up to in the garden and how good they are at their jobs. I was personally delighted to give up the chore of trying to eliminate the destructive voles that lived in my garden. I turned this job over to the black racer snakes that now reside in my garden rock pile. They are far better at the job then I was, and I don't have to lift a finger.

Before you dive into looking at creatures and insects in the garden, you must have a good resource to identify what is what, both the good and the bad. Do not do any handpicking, water blasting, or treating with any product until you have a confirmed identification. Sometimes it can be surprising.

OPPOSITE: *The power of insect-eating birds in the garden is underestimated. This goldfinch eats many seeds—but also consumes vast numbers of insects.*

Pollinators

Did you know that the pollination of plants is actually a side effect of a pollinator foraging for and collecting food? They come for the pollen and nectar, and as they go gathering from flower to flower, pollination happens. While some pollinators are more efficient than others, each one is necessary and has a job to do in the big picture of the world's landscape, food crops, and home gardens.

While the pollination services of bees and others were free and easy, few noticed them and all the good they did for the food and plant world. Then they started disappearing. In recent years, even home vegetable gardeners have begun to feel the impact of declining native pollinator populations. When the pollinators are abuzz and plentiful in a garden, the harvest is increased and the size and quality of the bounty is improved. Pollinators are necessary to grow a fruitful garden.

As gardeners are noticing a decline, many are acting to pump up the pollinator presence in the home garden. These actions will trickle out to help the extended population beyond the backyard. The way to attract all kinds of pollinating workers to the garden is to provide and protect habitat and to grow flowers in abundance from spring until frost.

During these years of gardening and farming, I have come to learn more about how to attract and support pollinators. Now I can clearly see why pollinators began to flock to my gardens as I was developing my cut-flower farm. Some of my practices were exactly what they were looking for. Planting large numbers of the same flower together helped the pollinators locate the flowers. In the home garden, it is beneficial to resist the urge to scatter flowers around the garden and instead plant them in a large mass. That offers the most benefit for pollinators.

These and other natural practices added over time have ushered in an abundance of pollinators to my gardens. Perhaps one of the greatest gifts of native pollinators is that they are right outside the back door. You need only invite them in and give them a few flowers, a spot to nest, and a habitat free of pesticides. The following are some favorite pollinators that visit my garden. The list is not meant to be all inclusive. To learn more about pollinators and supporting them, see the Resources section.

OPPOSITE: *The consistent presence of flowers in the garden will have the garden full of bumblebees. This genus of bees is on the decline and needs our help.*

My garden is loaded with butterflies. Providing a mix of flowers and host plants gives them flowers to feed from and plants to host their babies.

BUTTERFLIES

Butterflies are welcome in any garden even if they aren't the most efficient pollinators. Butterflies don't sport the pollen pouches of bees, and those long legs keep their bodies from brushing up against the pollen to gather and carry it off to the next flower. But none of this seems to matter as they float around the garden, going from flower to flower drinking the rich nectar to fuel their flight. Our garden hosts many butterflies, including swallowtails, monarchs, painted ladies, skippers, and others. They prefer flat landing surfaces with clusters of flowers. Butterflies taste with their feet! In addition to nectar flowers, it is important to include host plants in the garden where butterflies will lay their eggs. Caterpillars can live and eat from these plants, boosting the population of butterflies in the garden and beyond. Some host plants in my garden are milkweed for monarchs, dill and carrot (false Queen Anne's lace) for swallowtails, and thistle for painted ladies.

HUMMINGBIRDS

Nothing is more delightful in the garden than hearing the hum of a hummingbird heading my way. They are frequent visitors to the zinnia patches and even visit the zinnias I am holding in my hand after harvest! According to the US Forest Service, hummingbirds are important pollinators of wildflowers. The average weight of the hummingbird is equal to three pennies. Their hearts pump 1,200 times a minute, and their wings flap 70 times per second! They spend only 10 to 15 percent of their time feeding; the rest is spent perching and digesting. In addition to eating nectar, they also eat small insects such as gnats and mosquitoes. Hummingbirds are attracted to flowers that are brightly colored, tubular in shape, and open during the day.

Hummingbirds are one of the delightful bonuses of the flower garden. Once familiar with the sound of those fluttering wings, you will spot them often.

MOTHS

Perhaps the most delightful moth in the garden is the hummingbird moth. It looks like a hummingbird, sounds like a hummingbird, and even hovers and eats like a hummingbird, but it's a moth. Its tongue is longer than its body! You may be familiar with the spectacular sphinx moth with its 4- to 6-inch wingspan. Although strongly disliked by most gardeners—because they start life as hornworm caterpillars that munch on your tomato plant!—they are necessary pollinators and make a nice meal for birds, bats, and small mammals. These moths are especially fond of the large, white, late-day-opening flowers of the moonflower vine.

The hummingbird moth is often mistaken for a bumblebee until you notice the proboscis, the long, straw-like tube unrolling from under its head.

NATIVE BEES

There are 20,000 species of bees in the world. Captive honeybees are the most well known for their managed services to pollinate food crops. However, they are native to Europe and only represent a very small portion of the bee species in the world. The other species of bees are those that a gardener can attract to the garden to benefit from their pollinating services. In North America alone, there are over 4,000 species of native bees.

Pollination is serious business, and it tumbles right down the food chain onto the consumer's plate. A third of the food we eat and three-quarters of the world's

plants cannot reproduce without the pollination of bees, whether native or domesticated. Human-generated ways of pollination are already underway in the commercial world to help boost production in light of the serious decline of natural pollinators. The result of this will be felt by all at the cash register when buying fruits and vegetables produced via human intervention in place of the bees' free services.

A community of native bees has come to call my garden home through the years. I believe they were attracted because I grow mass plantings of flowers, and

I spot an abundance of different types of native bees in our garden. This little fellow is obviously different with his long antennas and short, stocky body.

some are their favorites. One of those favorites is the black-eyed Susan family. By including several different varieties of black-eyed Susans and varying the planting times, I have them blooming from late spring right through the summer season until fall. In addition, because the bees don't encounter pesticides in my gardens and I offer nesting options, the community continues to grow. It wasn't my intention to attract and provide for native bees, but to have the long-lasting cut flowers of black-eyed Susans available for as long as possible throughout the season. The wonderful result has been the abundance of native bees now living here.

WHERE DO NATIVE BEES LIVE?

Of the 4,000 species of native bees in North America, about 30 percent of them are solitary twig-and-wood nesters, and the other 70 percent are solitary ground nesters. The exception is the social bumblebee, which accounts for only 45 species. "Solitary" means that a female bee constructs the nest and provides food for the larvae alone, and "social" means that there is a queen bee that has workers to feed larvae.

The solitary wood-nester bee nests inside hollow tunnels in wood. These can be twigs of plants that a wood-boring beetle larva left behind or twigs with soft, pithy centers. It can also bore into soft, rotten logs or stumps aboveground.

Most native bees nest alone, not in a hive. Many nest in abandoned tunnels made by beetles and others in dead and dying trees.

The solitary ground nester excavates its nest below-ground. It burrows into the soil in narrow tunnels that have small chambers. These chambers are home to the brood cells that will be next year's bees. The tunnels can be as shallow as 6 inches and as deep as 36 inches. These bees must have direct access to bare soil to nest and often do it on gently sloping, well-drained land.

Social-bumblebees' nests can be above- or below-ground. They will use abandoned rodent burrows, bird's nests, and other small cavities. They will also burrow under plant debris. The queen lays eggs, and by mid-summer, the colony has many workers that visit nearby vegetables and flowers. The bumblebee is hailed as one of the most effective pollinators and the star pollinator of the tomato plant.

BEE FACTS

Learning what bees do for the environment and our food supply can help us understand why it is so important to help them not only to survive, but to thrive in our backyard gardens.

- Native bees are the most important group of pollinators in North America.

- Native bees are easily attracted to the garden with flowers and nesting sites.

- Native bees play a key role in a healthy environment.

- The majority of native bees live a solitary life alone, not in a colony or nest with others as the honeybee does.

A Case of Mistaken Identity

Solitary ground-nesting bees should not be confused with the yellow jackets that ground-nest together in great numbers. Yellow jackets defend their nests by attacking when the nest is disturbed by a lawnmower or threatened in some other way. These yellow jackets are often called "ground bees," which is a case of mistaken identification. They are not bees at all but wasps. It is documented and has been my experience that native bees are more agreeable in the garden because they don't have a nest of workers to defend. Don't let this false image of yellow jackets prevent you from encouraging native bees in the garden.

The yellow-jacket wasp has given the wasp and bee families a bad reputation. Most bees and wasps do not have the nest-guarding instincts of these guys.

Including perennials like heliopsis 'Summer Sun' in the landscape for endless blooms all season will provide for native bees and butterflies.

- Bees sting most often in an effort to defend their colony or nest. The majority of native bees aren't likely to sting except as a last resort when threatened, because they have no colony to defend. The social bumblebee is the exception, with a communal nest to defend. Of the 45 species, however, only 4 of them are considered feisty and easily provoked.

- A bumblebee's buzzing noise is the sound of its wings fluttering 11.4 times per second.

- Unlike the honeybee, native bees work even when it is cool and wet.

- Natives are known to start work earlier and work later than the honeybee.

- Native bees are valuable crop pollinators.

- Native bees are responsible for pollinating wild landscapes and sustaining native ecosystems.

- Honeybees make up less than 1 percent of all bee species yet receive the most headlines.

- The bumblebee is the strongest pollinator of tomatoes.

- Native bees come in all sizes, from the tiny stingless bee at .08 inches to the big leafcutter bee at 1.54 inches.

- The most common native bee, the sweat bee, is so small that it is often mistaken for a fly or wasp.

PROTECTING HABITAT

The natural first step is to protect any existing or potential nesting opportunities for native bees in your garden. This would mean keeping any dead wood or dying trees if they are not a threat to structures or people. Old dead wood is often full of leftover tunnels made by wood-boring beetle larvae. Wood-nesting bees will make good use of them.

Protecting areas of bare soil that are well drained and sparse of plants is another step to take. They make excellent potential ground-nesting sites. Ground-nesting bees have been found in urban lawns, landscapes, and orchards and alongside country roads. Even small spots of bare soil will be used by native bees. Permanent flower-garden pathways that have hard-packed soil are prime locations since flowers are close by. Mounds of soil left undisturbed are another good potential nesting site.

Providing water, nesting sites, and plants for pollen and nectar will keep the native bees in your garden. They will have no reason to waste energy traveling far to meet their needs. They won't need to venture beyond your garden to forage and risk an encounter with pesticides. They can put their energy into their offspring, not into searching for resources.

A glimpse of the native border that runs the perimeter of my farm and sustains many of the beneficial creatures that patrol my gardens.

Most native bees go about their business of pollinating, nesting, and reproducing without being noticed. Many are so tiny we don't even think they are bees.

PROVIDING HABITAT

There are several easy steps you can take to attract native bees to your garden. Keep in mind that it may take a year or more before the bees find and begin to use them. Each of the following nest projects can be done and placed individually, or they can be incorporated together in a beneficial hotel as we have done (see page 116).

Make a Sand Pit

Ground-nesting bees will nest in well-packed sand. In a well-drained, sunny spot, dig a 2-foot-by-2-foot hole. Fill with fine-grained sand or sandy loam. Tamp the sand to compact it slightly. Excellent drainage is necessary. If the hole fills with water, it will kill the nesting bees. In areas with poor drainage, a large container can be used, or the sand can be piled aboveground.

Twig Nesters

Planting shrubs and perennial plants that have dried hollow stems at the end of the season is a way to make good nesting sites for twig-nesting bees. Don't remove the twigs from the plant; leave them for the bees to use the following growing season. I find that my hydrangeas provide such an abundance of hollow stems from boring beetle larvae that some can also be broken off and added to our diverse beneficial hotel.

A Case of Mistaken Identity

I often show a photo of this insect in my lectures. I ask for a show of hands of those who recognize what insect it is. I am saddened to say that less than 5 percent of the attendees raise their hands. Considering that I am often speaking to seasoned gardener groups, it is surprising. Jumping into action without knowing what is in the garden and what they are doing can lead us astray.

Do you know what insect this is?

Go to pages 123–124 to learn what this is and what its job is.

Leaving the old hollow stems from last year's crop of hydrangea blooms can provide nesting sites for native bees. Birds also use them for perches.

Beneficial Creature Hotel

This multilayer structure is made from recycled shipping pallets and other items. It can provide habitat for a diverse community of creatures, from native bees and beneficial insects to birds and small mammals. It is a place to rest, hide, and forage for food in addition to nesting.

To add a layer of cover for wildlife, I place the first pallet on a base of six cinder blocks to allow birds and small mammals access underneath. It is a full house when it is pouring rain! Additional cinderblocks are useful to add height. Bricks have the holes aligned to fill with twig bundles. Other materials include landscape cloth for the planter; potting soil and low-growing drought-tolerant plants for the top (sedums are excellent); and moss, pinecones, twigs, sheets of bark, straw, small logs, and other natural debris to stuff the pallets.

Place the hotel in a level, sunny location with protection from north-northwest winter winds. A row of evergreen shrubs or a wood fence placed 10 or more feet from the hotel offers a good windbreak.

1. Gather five to seven shipping pallets. Look for pallets with either an HT (heat treated) or a KD (kiln dried) stamp, which indicates no chemicals were used. *Do not use* pallets with the stamp MB (methyl bromide, a broad-spectrum pesticide).

2. Place six cinderblocks with the holes facing outward on the ground and top with a pallet. Use a level to be sure the site is flat. As you build it up, an uneven base will become more obvious and the hotel will be unstable.

3. Place six more cinder blocks on the first pallet with holes facing out and top with a pallet.

4. Continue to stack three to four more pallets one on top of another.

5. Flip the top pallet to expose the wide slat cavities that work as planters.

6. To create the planter, line the openings with the woven landscape cloth and secure with staples. Cut a few slits in the landscape cloth for drainage. Fill the planter with soil and plant low-growing plants. Place a few rocks among the plants for insects and reptiles.

7. Fill the holes of the pallets and cinder blocks with the nesting materials gathered.

Our beneficial creature hotel has to be restocked from time to time. The birds take dog hair, garden twine, and small sticks for nest building.

Make a Bee Block

Cavity-nesting bees will move into blocks of wood that have varying depths of smooth holes drilled into them.

- Use untreated wood.

- A 4-inch-by-8-inch post is ideal, making the block at least 8 inches tall. Logs can also be used.

- The drilled holes should be smooth inside, and the ends should be closed.

- Drill holes of ⅜ inch or less in diameter.

- Holes less than ¼ inch in diameter should be 3 to 4 inches in depth, and those over ¼ inch should be 5 to 6 inches in depth.

- Place the block with the entrance holes facing south-southeast to warm the nest on cool days. Placing it on a post or in a tree works well. Keep it off the ground.

- Making a roof for the block will help prevent rain from soaking the block.

About 30 percent of native bee nests are in dead wood or somewhere similar. We have included bee blocks in our beneficial creature hotel for them.

Stem Bundles

- Bundle 15 to 20 stems or straws together.

- Some plants produce naturally hollow stems, such as bamboo, elderberry, reeds, yucca, and lunaria.

- Cut the stems at a node so that one end will be closed. A drop of caulk at the end can also serve to close the stem.

- Stems should be 6 to 9 inches long.

- Cardboard straws are available for sale if stems are not available. Be sure to keep them dry.

- Cable-tie the bundle together.

- PVC pipe can also be filled with stems or straws.

- Place the bundle in a spot protected from the worst of weather with the entrance holes facing south-southeast, where it will benefit from the early-morning sun.

- The bundle should be horizontal and can be placed at any height. Secure so it won't blow in the wind. It can be attached to a building, fence, garden stake, or tree.

A stem bundle of bamboo hanging in a tree bordering our garden. The stems are open on one end and closed on the other.

Beneficial Predators

You may meet some predators in this chapter that are surprisingly beneficial. A few of these were not welcomed into my garden during those first years as I was finding my way back to nature. Some of these creatures terrified me (wasps). In the case of others, such as hawks, I was afraid of what they might do to some of the other creatures I liked in the garden. It took me time to realize that some of the very garden problems I was suffering from were a direct result of the lack of natural predators.

After years of doing my best to restore the natural order of my garden, I have learned this: the cycle of nature works, and I must give it what it needs and trust it to work. I have stood by and watched as a hawk carried off one of our prized black snakes. Snakes control the voles in my garden, but I quickly forgave the hawk because its other main diet selections are mice and rats, both of which I am never sorry to see go. Sometimes the food chain goes both ways— some beneficial creatures become food for other beneficials. I have learned to live with this because it is how nature works.

The snapshot of nature is big and wide. A recent discovery for me was that possums eat ticks! Possums are creatures that we are all pretty hard on because they are just not cute or pretty. According to the scientists of the Cary Institute of Ecosystem Studies, possums can kill 5,000 ticks in one season. They are the only marsupials in the United States, and they were waddling around in the days of the dinosaur. Cuteness no longer seems so important to me.

Keeping our minds, ears, and eyes open to what is going on in the garden and making the most of all who venture in will help grow a healthier garden. When we provide the groundwork—a diverse garden—they will all come of their own accord.

OPPOSITE: *Wasps prey on insects to feed their meat-eating young. I count them as some of the most hardworking beneficial insects in my garden.*

I watched as countless insects were fed to these baby house wrens. This house is conveniently located right outside my back door and is a regular source of entertainment.

BIRDS

Birds contribute far more to a garden and landscape than the obvious pastime of birdwatching. They eat a significant number of insects because insects are the bulk of many birds' diets. One has only to sit and watch birds flitting about in a garden to soon realize they aren't just goofing off out there—they are hunting for bugs. While the volume of insects required to sustain an adult bird is impressive, it's when they start feeding a brood of babies that their hunting skills get pushed into overdrive.

A sparrow perching atop the curly willow watching for its next meal. This tree and shrub border around the garden is full of bird activity.

The robin is a constant companion in my garden. Robins consume a significant number of caterpillars and grasshoppers' which makes them always welcome.

Many birds can be attracted to the garden with water and feeders. If you install nesting boxes to suit birds native to your region, from songbirds to owls, they will take it as an invitation to move in and raise a family of hungry babies safe from predators. Meet some of the birds of my garden:

BLUEBIRDS

I spent years trying to convince eastern bluebirds to move into my garden. Little did I know that providing water would be the ticket. At first they came to bathe, and then once nesting boxes made to bluebird specifications were installed, they finally moved in. Watching these birds catch insects to deliver to their mates and their babies is one of my favorite garden pastimes. I've recently learned of vineyard owners in the west who install nesting boxes among the grapevines to attract the western bluebirds for their insect control. Bluebirds eat grasshoppers, crickets, beetles, and moths.

CHICKADEES

We have a nesting box in the shade garden right outside my back door. Each spring, a tiny chickadee builds a beautiful nest with the moss from my garden and raises a family. In summer, 70 percent of the chickadee's diet is insects. They hide and store food for later and can remember thousands of hiding places. Chickadees eat aphids, whiteflies, caterpillars, ants, earwigs, and other insects.

AMERICAN ROBINS

We have an abundance of robins running up and down our pathways and perching on garden stakes looking for their next meal. They are especially fond of the birdbaths. Robins' diets include beetle grubs, caterpillars, grasshoppers, and earthworms. They also eat berries, with strawberries being a favorite. A row cover on the strawberry patch prevents the birds from helping themselves.

TUFTED TITMICE

These little insect-eating birds line the inner cups of their nests with hair, sometimes from live animals. I once watched as one repeatedly visited my golden retriever lounging on the back porch. The bird would pluck a beak full of fur and fly off with it, only to return in a few minutes for more. Tufted titmice eat caterpillars, beetles, ants, wasps, stink bugs, spiders, and snails.

BARN SWALLOWS

Each evening as the sun starts to drop in the sky, I welcome the chatter of these birds as they put on a show flying over my garden. They like open space and a fence line to land on. As they fly back and forth and up and down, they're gobbling up some of my peskiest insects: mosquitoes, gnats, and flying termites.

This Cooper's hawk is a frequent visitor to our garden. He has helped to reduce our vole population; thus, our plant loss to voles is almost non-existent.

States, a region home to over 100 species of reptiles and amphibians not found anywhere else in the world. I am painfully aware of this group and their loss of habitat. My deep concern arises as a neighboring farm is currently under development. Its wetlands have hosted a wealth of amphibians and reptiles.

Many of these guys don't show their faces as boldly in the garden as some other creatures, but be assured that each plays a key role in the ecosystem of our gardens and the surrounding wild landscapes. You might think of this group as the indicator of the good and bad the environment has to offer. This is because they often live in or around water and depend on it for their existence. This water home base can be a haven of life or a cesspool of runoff pollution.

RAPTORS

If you face damage from voles (garden mice), rats, gophers, moles, or other rodent pests, perhaps you should consider inviting some of their natural predators. Hawks, owls, and other birds of prey are nature's fix for these highly damaging pests. When I hear the hooting of an owl, I practically jump for joy! A small screech owl under 10 inches tall consumes many voles, mice, and insects from a garden, and when they are feeding babies, the number skyrockets. Farmers consider red-tailed hawks to be some of their most useful residents since 80 percent of their diet is mice, rats, and the occasional snake. Providing adequate cover for songbirds and other possible targets can help to protect them from falling victim to birds of prey.

ATTRACTING AMPHIBIANS AND REPTILES

- Add a pool or pond to the garden, placing it so the creatures have safe access to and from it.

- Include native plants in your yard.

- Provide shelter such as dead trees and brush, leaf, and rock piles.

- Keep pets such as dogs and cats out of the habitat area.

- Say no to all harmful chemical use.

SNAKES

I'm not a fan of snakes, but I have learned to tolerate them because I am even less a fan of voles. It's easy to see why black racer snakes have taken up residence in my garden—it's because I have a healthy population of voles. Or at least I used to until these guys moved in. You don't normally see voles running around aboveground in the garden, just the resulting damage they do as they sneak around underground and eat flower bulbs, root vegetables, and the roots of plants, shrubs, and trees. The number-one cause of loss on our farm

REPTILES AND AMPHIBIANS

One of the signs of a healthy habitat is the presence of frogs, lizards, turtles, and, yes, even snakes. Their absence may indicate missing habitat features or an environmental problem. I live in the southeastern United

The black racer snake eats rodents like voles and mice. I'm not a fan of snakes, but they do such good work I've learned to live with them.

used to be these rodents. I spent years trying to eliminate them with little to no success. Voles were a constant problem until I added a pile of rocks and some favorable places for these snakes, which are harmless as long as you aren't a vole or mouse.

FROGS

I find a variety of frogs in and around the gardens. The green treefrog is so frequently a visitor to the outdoor work area where we store our green harvest buckets that it has become commonplace to check each bucket for a frog before use. Most frogs eat a steady supply of insects, spiders, snails, and worms. This makes them more than welcome anywhere they'd like to take up residence around here.

BENEFICIAL INSECTS

There are many helpful insects commonly found in the garden. Here are just a few.

LADYBUGS OR LADY BEETLES

Certainly the poster child of the good bug world, the adult ladybug is cute and well known, but the ladybug larva is not. This is a perfect example of why you must

A ladybug in hot pursuit of the common garden pest, the aphid. A ladybug can eat up to 50 aphids a day.

identify every insect you find in the garden before proceeding. The larva of the ladybug is often mistaken for a pest and is eliminated by gardeners. It is so important to know what you have found before you decide to do anything about it.

Ladybugs and their larvae are efficient consumers of aphids in the garden. Aphids are a common garden pest that causes puckered foliage, among other troubles. I find ladybugs in all stages of life all over our garden. I have been known to help some of our ladybug larvae

Ladybug larvae consume a vast number of aphids. They tend to stay put as pests get scarce, while the adults fly off. PHOTO BY GARDENER'S WORKSHOP FARM

relocate from one area of the garden to another where there is an outbreak of aphids. I use a sheet of paper to gently lift them off. After gathering several, I walk them over to the infested plants, where they'll have a plentiful supply of food waiting for the next few days.

WASPS

If you want some serious caterpillar and insect control in your garden, you better leave your fear of wasps behind like I did years ago. The wasp is a very important part of natural pest control in a garden. Plant and blossom damage from various caterpillars has all but vanished from my garden since I gained a healthy respect for wasps. I have let them have their way around my farm where it doesn't interfere with everyday human existence. I often find a concentration of wasps in the marigold patch, where they seem to be fond of backing into the blooms to lie in wait for prey.

There are thousands of species of wasps worldwide. Each species either lives together with a queen and workers or lives alone, breeding and making nests independently. All are beneficial to the garden. A very active

wasp in my garden is one that captures caterpillars, beetle larvae, flies, and other insects to sting and carry back to its nest of meat-eating larvae. Yes, baby wasps are meat eaters, and that meat comes by way of insects from the garden. Interestingly, when the insect is captured, the wasp stings to paralyze the insect, not to kill it. This prevents the insect from rotting before the larvae has a chance to eat it.

Other wasps lay their eggs right on the actual pest insects in the garden. This group, known as parasitoid wasps, are held in high regard for their ability to lay eggs on pests. When the larvae hatch, they consume and kill the pest. These wasps are generally pretty tiny, many only $\frac{1}{64}$ to $\frac{5}{16}$ of an inch long. If you ever spot a giant green hornworm eating a tomato and it is covered in white cocoons, leave it be! Those white cocoons are the larvae of the braconid wasp, and the hornworm is their host and home. As the wasp larvae develop, they will consume and kill the hornworm and then grow up to lay their eggs on other hornworms.

While many adult wasps feed on nectar, they spend most of their time gathering insects for their young.

SPIDERS

Spiders have perhaps the creepiest reputations. They, like the wasp, are some of the most ferocious bad-bug consumers in a garden. Unfortunately, their quick movement, long legs, and fuzzy bodies have put them on most homeowners' hit-and-squish list. Virtually all spiders feed on insects as well as other spiders. The webs they spin in the garden are excellent traps set for any number of insects—sometimes good bugs get caught there, but for the most part, I find stinkbugs, grasshoppers, beetles, flies, crickets, and other insects. Spiders can be found scurrying over our beds and weaving webs all over the garden. Anything that helps eliminate a stinkbug from my garden is a friend of mine.

SOLDIER BEETLES

Sometimes called a leatherwing, the soldier beetle resembles a lightning bug without the organ necessary to light up. The larvae of the soldier beetle have an unquenchable appetite for soft-bodied insects and will even devour those much larger than they are. We have a strong presence of soldier beetles since the adults favor pollen and nectar. Zinnias, marigolds, and the fall-blooming goldenrod are some of their favorites.

A garden spider that has trapped and wrapped insects up for a later meal. I often see stinkbugs in these spider lunchboxes.

An adult soldier beetle on a strawflower. These insects feed on aphids and their larvae, on the eggs and larvae of grasshoppers and on other garden pests.

Dragonflies are ferocious predators of other flying insects, with the mosquito being a favorite. Their larvae live in water and will eat anything smaller than themselves.

DRAGONFLIES

These guys are aggressive predators to insects and live a fascinating life. They have the rare and stellar ability to capture insects, such as mosquitoes, in flight. They also have what we would consider superpowers in vision and are hungry all the time. The larvae of dragonflies live underwater and can do so for up to 5 years, all the while eating mosquito larvae and other small insects. Attract dragonflies to the garden with convenient landing posts such as the garden stakes that hold up flower-support netting and trellising.

ANTS

What hard workers ants are in the garden! Most of their jobs go unnoticed or even misjudged. Take their presence on peony blooms. The ants are drawn to the sticky, rich food source on the buds, and they in turn patrol and protect the plants from predators. To remove the ants from a cut-flower peony stem, place the stem into water with cut-flower food, and the ants will be drawn to the sugars in the water and vacate the stem. Ants are also like little mini-rototillers, rooting around and moving soil in the same way earthworms do. They are key players in dispersing seeds throughout forest and woodlands. Leave ant mounds undisturbed when possible.

SYRPHIDS OR HOVERFLIES

The larvae of the hoverfly can consume up to 60 aphids a day and also eat thrips, one of the flower gardener's toughest-to-control pests. Some adult flies are important pollinators of flowering plants worldwide. This stellar fly hovers over the garden to spot a colony of desired pests. Once located, it touches down and lays eggs. The larvae emerge and consume the pests. Dill and false Queen Anne's lace (*Ammi majus*) are favorites of the hoverfly.

GROUND BEETLES

Most ground beetles are shiny black or metallic. They normally do their hunting at night and are predators of caterpillars, snails, slugs, and other soil-dwelling insects. They are fast moving and can have enlarged mandibles for grasping prey. Brush, dead wood, and leaf piles are favorite daytime hangouts for this group.

The ground beetle gets busy at night consuming slugs, snails, caterpillars, and other pests. Ground beetle larvae live below ground and also devour soil-dwelling pests.
PHOTO BY JEFF HOLCOMBE, SHUTTERSTOCK

ASSASSIN BUGS

A pack of assassin-bug nymphs was my first experience with beneficial insects in the garden. After watching a group of nymphs carry off a caterpillar, I was hooked on putting nature to work in my garden. This family is large, and they aren't picky about what they eat, including both bad and good insects. They will wait patiently for an insect to come along or stalk them and take them down. I find them wherever other insects are present in the garden and on more than one occasion have come upon them eating stinkbugs. Assassin bugs can bite humans, which is painful but not considered dangerous.

The assassin bug is pictured here eating a tarnished plant bug—one of my worst enemies. Assassin bugs also prey on stink bugs, Japanese beetles, and the Colorado potato beetle. PHOTO BY PETER ELLSWORTH, USDA

SECTION 4

Growing a Healthy Garden

As I write this section of the book, my garden is 6 weeks into the growing season. While this is proving to be a challenge as the book and the farm jockey for every minute of my time, I am reminded on each trip to the garden why I do and don't do certain chores. What has become second nature for me is being highlighted just in time to share here.

I will confess now that there are still times I don't follow my own suggestions and pay the gardening price. This was evident as I walked out to the bush-pea patch just weeks ago to find the entire crop of beautiful 3-inch pea shoots pulled out and left shriveled and dead. Undoubtedly they were pulled out by a crow looking for the pea seed to eat. This type of damage is preventable by simply covering the plantings with a row cover. Even those who know what to do to prevent problems sometimes just don't do it, and I am continuing to live and learn hard lessons in my own garden.

Beyond offering that I am an urban cut-flower farmer growing in the midst of a city in southeastern Virginia, I haven't shared my city growing conditions and challenges. On the 2½-acre property that includes my home, I have 1½ acres of working gardens. These gardens are flanked by permanent planting areas. I do not have a greenhouse or any hoophouse structures. All of my flowers and vegetables are grown outdoors in a garden. I do all of my seed-starting indoors in a 10-foot-by-10-foot room equipped with seedling heat mats and shelving units with hanging grow lights. During the peak years on my farm, we started 100,000 seedlings throughout each year and harvested over 10,000 stems of flowers each week in season. It is possible to grow a lot in a small space and do so with little infrastructure.

The permanent plantings that surround the gardens have proven to be a huge benefit to the overall health of the gardens. A few years ago, I planted a 12,000-square-foot native tree, shrub, and perennial border. This hedgerow is now home to a beneficial community of creatures here on the farm and helps them stay put.

Growing a healthy garden is good for you, the environment, and all the creatures that live in it. I've learned that growing naturally strong plants, surrounding them with as many native plants as possible, and using common-sense gardening practices have created this gardening good life—it's just as a garden should be.

OPPOSITE: *My farm is safe for people, pets, and wildlife to sit, live, and eat. Here is Beri modeling a feverfew flower crown in a field of buckwheat.*

Tending the Healthy Garden

The road to a healthy garden may be a little bumpier for some gardeners than others. We all start with different conditions and have different expectations of what is tolerable. I am gardening in the midst of the city. While I have great soil, I also have fairly strong deer pressure but can't install deer fencing because of city ordinances. Fourteen deer in my garden at once is the record so far. Our squirrels have so little habitat left in the city that they can be a force to reckon with as they carry off our produce and flowers to feed their family. What works in my garden may not work in yours, but the overall mindset to tackle problems in the garden in a new way remains the same. Set the garden up to succeed, be patient, and know that it is possible.

The root of the solution is preventing problems. If a problem does occur, consider whether it was preventable and can be resolved without resorting to a pesticide intervention, organic or otherwise. Making choices about what to grow and what not to grow can have a big impact in creating or preventing potential problems in the garden. There are some flowers, such as gladiolus and Shasta daisies, that I simply do not grow anymore because they are pest magnets in my garden. Both attract and do a fine job of harboring thrips, a major flower-garden pest. This is also true for butternut squash, which requires a long growing season. I have found it nearly impossible to control the squash bugs over such a long period. What was my solution in these instances? Pull the plants, bugs and all, and send them packing to the trash truck. I no longer grow these plants because there are far too many beautiful, problem-free flowers and delicious vegetables to grow instead. Why invite problem plants and their troubles into my garden? If you have a pesky, needy plant, dig it up and throw it away!

In addition to growing naturally strong and healthy plants in my garden, I focus on getting the most out of my garden space. If it is possible to get the abundance of a large garden from one smaller, why not go for the smaller size? This creates a garden that naturally has less maintenance and more bounty. When common-sense gardening practices are applied along with succession planting, you will reap the greatest bounty of flowers, vegetables, and herbs from the least amount of space.

Taking extra time to set up the garden at the beginning of each season will have a huge payoff throughout the growing season. It wasn't always this way for me. I remember spending more time weeding the pathways of my garden than tending the crops I was growing. Don't fall into this trap! I'm always sorry to find myself there. This season, we ran short of leaf-litter mulch and tried to make a little go too far in the pathways. The pathways with deep mulch were gorgeous and weed free, while those with the thinner mulch had weeds growing at the speed of light. Plan to prevent summer weeds and then follow through. I am afraid that many a potential gardener never got a good start in gardening because summer weeds took them down and ran them off. The garden is much more enjoyable with few to no summer weeds.

Preventing labor-intensive chores such as weeding and the grief of damage from pests can go a long way in making a garden more successful and enjoyable. The following are some gardening practices that have led me to an abundant, pesticide-free garden.

A view of my garden in the midst of the summer season ready for harvest: beds of sunflower 'ProCut', zinnia 'Benary's Giant', and heirloom tomatoes.

My garden beds are grouped together and surrounded by pathways covered in various types of mulch. This minimizes maintenance during the growing season. PHOTO BY GARDENER'S WORKSHOP FARM

LOCATING THE GARDEN

A cutting and vegetable garden should be located in full sun. In my experience, 8 or more hours of full sun daily is best for a garden to thrive and reproduce. A lack of sunlight will lead to puny plants. They will grow tall and leggy as they stretch to find more sunlight and will not produce.

Take time to think through what will surround the garden. Will it be lawn or pathways with mulch or some other covering? Thinking these options through before you start could save a season of chores and aggravation.

My beds are grouped together in an area with pathways that surround those beds. The pathways are finished off in different ways depending on the time of year the beds are prepared and what I have on hand to use. The lowest-maintenance pathways are free of all vegetation and are covered after bed preparation is complete. My goal is to create an area where the pathways need no maintenance and the garden's edge can be easily mowed and kept tidy. Creating a garden with permanent beds and pathways is an excellent way to lower maintenance.

Soil testing is easy, inexpensive, and necessary. It provides a look into the science of your soil by professionals that can guide you on any additives needed.

GARDEN SOIL

Much of the focus on my farm is on feeding and protecting the most precious gem we have been given: the soil. What the plants sink their roots into is perhaps the most significant contributor to their overall survival and health. My goal is to give the soil what it needs to be as self-sustaining and healthy as possible.

SOIL TESTS

Every fall, I soil-test our gardens. This brings to light any deficiencies or excessive values in our soil. Doing this in the fall allows time for any corrections to be made. Recommended supplements can be applied and allowed to mellow over winter. If you are not soil testing, you are gardening with a blindfold on. Testing is easy and inexpensive and can be revealing. It is a garden report card, really. I do not add anything to our soil beyond organic matter and organic fertilizer unless a soil test tells me to.

FEEDING THE SOIL

Adding organic matter to the soil must be done for the life of the garden; it's not a onetime application. This organic matter is what feeds and maintains what I consider to be the most important players on the farm: the microorganisms that live in the soil. These guys process and deliver the nutrients to the plants as one of their many important jobs. In gardens with long, hot growing seasons like mine, the goal is to add 2 to 4 inches of some type of organic matter to the garden each year. In gardens with shorter, cooler growing seasons, 1 to 2 inches may be sufficient. The longer and hotter the growing season, the more active and hungry the microorganisms in the soil are. After taking such good care of the garden soil, take care not to walk on the beds. This will compact the soil.

ORGANIC FERTILIZERS

We use dry and liquid organic fertilizers. For watering transplants, add organic liquid fertilizer to the watering can once a week per product instructions. When preparing the planting bed, incorporate dry organic fertilizer into the soil per product instructions. Plants will really appreciate a monthly soil drench with the liquid fertilizer and a foliar feeding once planted in the garden. A soil drench is pouring liquid fertilizer mix onto the soil or running it through an irrigation system. Foliar feeding is to sprinkle or spray the liquid fertilizer mix on the leaves. I stop foliar feeding when buds or fruit begin to form because the fertilizer is aromatic, and you won't want your cut flowers or vegetables to have this smell. I like to use fertilizers that are made from sustainable products such as seaweed, fish, and chicken litter.

Leaf litter is always the preferred mulch in my garden pathways. It adds tons of organic matter, retains moisture, and harbors beneficial creatures.

PATHWAY COVERING

I have found that maintaining pathways with organic matter keeps my garden soil cooler, feeds the soil, retains more moisture, provides habitat for beneficial soil-dwelling creatures, and generally leads to a thriving garden. There are exceptions and situations where I use an alternative method. There are sections of my garden beds that are torn down each season and rebuilt, while other sections are permanent gardens. I apply these methods in both.

MULCHING

When leaves are available, they are my first choice. All pathways are mulched with 6 to 10 inches of leaves with a 2- to 3-foot border of the same around the garden perimeter. We dump thousands of bags of leaves collected from curbsides in nearby neighborhoods. We do not chop the leaves before using them. This mulch retains moisture, suppresses weeds, and keeps the soil cooler. Deep perimeter mulch has proven to be the best strategy for stopping creeping grasses and weeds. Mowing around the perimeter with a tire in the mulch is quick and easy. Other organic mulches, such as bark or chips, will work but are costly to apply at such a depth to

prevent weeds. I have not found slugs and snails to be a problem when using leaf litter. I believe that a strong presence of birds and other slug-eating predators in my garden keeps them in check. Using a layer of cardboard under the mulch has been one of the best methods to smother and kill persistent perennial weeds.

LANDSCAPE CLOTH

I use woven landscape cloth for just about everything except what it was intended for. It is an excellent choice for pathways or garden perimeters where leaf litter isn't suitable or available.

- Beds that will have seeds directly planted in them have bare soil until the seeds sprout and have grown enough to be mulched. Covering the pathways by these beds with leaves would make the weekly hoeing of the bed tops troublesome. The landscape cloth provides excellent pathway weed control and kills the weed seeds on the soil's surface under the cloth. I use 36-inch-wide woven cloth (not the felt-backed cloth.) Make the pathways 24 inches wide, leaving 6 inches of fabric to cover the sides of the raised beds. Use a rubber mallet to drive 6- to 8-inch-long metal staples every 4 to 8 feet along the pathway to hold the cloth down. At the ends of the pathway, drive the staples closer to secure the ends. I do not leave the landscape cloth down for more than

I use white biodegradable film on beds to prevent overheating young transplants in the summer heat. Landscape cloth in the pathways is for weed prevention.

a growing season. Creeping perennial weeds can jump on top of the cloth and run or even grow through the cloth, becoming well established and impossible to remove.

- Gardens using lay-on-the-soil surface irrigation will benefit from a 36-inch-wide piece of landscape cloth running across the head of the garden, butting up to the heads of the beds for the irrigation system to lie on. This prevents vegetation from growing up and wrapping around the system.

- See the weed-prevention section (page 147) for other uses of landscape cloth.

GROWING PATHWAYS

Allowing natural vegetation (grass and weeds) to grow in the pathways is a choice I go with when I don't have any other options. While it works, it has to be mowed at least weekly or the pathways are hardly passable for garden chores. The real downfall in addition to mowing is that the vegetation will creep into the beds, creating more chores.

A 5-by-7-inch tray holding 40 ¾-inch mini–soil blocks with newly sprouted zinnias. They will grow for 2 weeks and then be planted in the garden.

STARTING THE GARDEN FROM SEED

Seed starting is a miraculous gardening experience. When you choose to start your own seeds, you can be assured of pesticide-free plants of endless variety available at the best planting time for your garden. The benefits of seed starting are worth the time it takes to become familiar with a few ground rules.

There are two ways to start seeds: indoors in a container or soil block and outdoors in the garden. Each way may have benefits for different gardens in different seasons. While I have great success planting seeds in the garden in fall, I have dismal results doing it in spring and summer. Gardeners in northern regions may not have a long enough growing season to plant seeds in the garden and get blooms or fruit before frost. They start indoors weeks before their planting date to have a more mature transplant that can bear fruit or blossoms sooner for their short season. Seed starting is not one size fits all; find what works best for you and your garden.

OUTDOOR SEED STARTING

One of the limiting factors on when to plant seeds outdoors is the temperature. Warm-season tender annuals require nighttime temperatures staying above 60 degrees Fahrenheit before planting seeds. Cool-season hardy annuals need cool nights below 60 degrees Fahrenheit with warm days for the best sprouting in the garden.

Outdoor Planting Tips

- Plant seeds the same day you prepare the bed or disturb the soil by running a hoe throughout the top 1 inch of soil just before planting. This will eliminate any weed seedlings that are not visible yet but began developing since the soil was last disturbed.

- Plant in a pattern that is easily marked. Straight lines are often used in cut-flower and vegetable gardening and make for easy seedling identification and weed-prevention chores. Use plastic picnic knives as plant markers. Place one at each end of the line of planted seeds.

- Plant the seeds into the bottom of a shallow *V* trough. Water only in the trough. This waters just the planted seeds, not the weed seeds all over the surface of the bed.

A young bed of bush beans. The seeds were planted directly into the garden once nighttime temperatures stayed above 60 degrees. PHOTO BY GARDENER'S WORKSHOP FARM

- Cover the seedbed with a floating row cover to prevent birds from eating the seeds and for increased and quicker sprouting (especially sunflower seeds).

- Push the floating row cover to the side when watering the seedbed to keep it evenly moist.

- Run a garden hoe throughout the bed weekly except where the seeds are planted to prevent weed seedlings from developing.

- Thin seedlings when they reach 3 to 5 inches to the recommended spacing per the seed packet.

- Remove the floating row cover once the plants are big enough to be mulched approximately 5 to 8 inches. No more weekly hoeing is necessary.

Planting warm-season seeds directly into holes in black biodegradable film in summer has proven to be easy and successful. Here, a cucumber seedling is emerging.

USE MULCH FILM FOR EASIER DIRECT SOWING

In recent years, we have begun sowing seeds into holes in the biodegradable film we use. This works well in the heat of summer for that second or third planting of warm-season annuals such as zinnias, sunflowers, squash, and cucumbers.

1. Prepare the bed and cover as described in the weed-prevention section on page 147.

2. Plant two to three seeds in each hole. The heat from the sun on the black film encourages and speeds the seeds' sprouting while the rest of the bed is covered and does not receive sunlight to sprout weed seeds.

3. Thin the seedlings to one plant per hole when they're a few inches tall.

INDOOR SEED STARTING

Starting seeds indoors allows a wider window of planting times and brings on earlier blooms and fruiting. This earlier start can also help beat insects to the fruit and blooms with planting timing.

I use the English seed-starting method called soil blocking to start seeds indoors. I was drawn to it 20 years ago by its incredible space savviness because I did not and still don't have a greenhouse. All of my seeds are started indoors in a home-like setting. Soon after I began using soil blocks, I started to tweak the method to suit an indoor tabletop setting. First, I began using flat-bottom trays with no drainage holes, which allows the trays to be watered indoors with no water leaks. As

I started growing more plants and began running out of room, I found that I could use the smallest blocker (¾ inch) for most seeds and that it could and would support a beautiful transplant until it was planted in the garden. The key is to plant it in a timely manner. I use the larger 2-inch blocker when a seed is too large for the small blocker or to move up a small blocked plant to into the 2-inch block to grow into a larger transplant. This tweaking led me to where I am today—starting tens of thousands of seedlings a year indoors in a heated and air-conditioned 10-foot-by-10-foot workroom with seedling heat mats and shelving units with grow lights. You can find suppliers in the Resources section.

Soil-Blocking Benefits

- Space savvy; 40 small blocks fit on a 5-inch-by-7-inch tray.

- High rate of germination.

- Transplants resist transplant shock naturally because the roots do not become potbound.

- Plants earlier to bloom and fruit.

- Requires a shorter indoor growing time because seedlings tend to grow faster and can be planted younger.

- Reusable trays are environmentally friendly.

It is an efficient method that suits a home setting nicely and produces a seedling that hits the ground running in the garden. Take a look at the photos at right for a tutorial.

Soil blocking is a tabletop seed-starting method convenient for in-home use. It also takes up little space.

1 *With the blocking mix wet, hold the soil blocker with two hands and push it into the wet blocking mix, filling the chambers.*

2 *Using a flat-bottom tray without drainage holes, depress the plunger and release the blocks onto the tray.*

3 *Dump the seeds into an aluminum pan that has no static electricity. Use a toothpick moistened with saliva to pick up and place a seed on each block.*

4 *To encourage more and faster germination, place the trays of blocks onto a seedling heat mat to warm the soil.*

5 *Once 50 percent of the blocks show signs of sprouting, move from the seedling heat mat to a grow light.*

Indoor Seed-Starting Tips

- Whenever possible, I always choose to start seeds indoors. It is easier and more reliable to start and tend seedlings indoors without the outdoor elements to contend with.

- Seeds need consistent warmth to sprout. A seedling heat mat warms the soil, which leads to more seeds sprouting more quickly. Once 50 percent of the seeds show signs of sprouting, move from the seedling heat mat to the light. Growing a healthy transplant indoors requires 16 hours of sunlight a day. A grow light turned on and off with a timer is the best way to provide it.

- When the seedlings have been moved under the lights, once a week I add organic liquid fertilizer to the watering can according to the fertilizer instructions.

- When seedlings reach 3 to 5 inches tall, move them outdoors to a protected area to harden off for at least 5 days before going out in the garden to face the elements. My seedlings sit under a carport that receives morning and late-afternoon sun with a small amount of wind.

- Cover the transplants with a floating row cover once planted to protect them from wind, rabbits, squirrels, birds, and deer. Leave the covers on for 10 to 14 days or until the plants are well established. I prefer to hoop and cover, which allows for longer use of covers and less threat of overheating. See the Floating Row Cover section on the next page for more information.

Move seedlings outdoors to a protected area once they grow to near the planting size of 3 to 5 inches, to acclimate them to outdoor weather.

PINCHING TRANSPLANTS FOR MORE BOUNTY

I will confess that I am not a very faithful pincher. This is a result of there being too many other demanding tasks on a farm. However, every time I do pinch, I am reminded of the benefits. Plants will produce more stems of better cutting quality when pinched than when I leave that first central stem to grow into a monster stem.

Which to pinch: Any *branching* annual, such as zinnias, marigolds, and basil, can be pinched. Pinching establishes branching that encourages more stems to grow and will also help to control the height of a plant. Keep in mind that pinching will delay blooming a bit, but the bonus is that the lower you pinch, the more stems you will ultimately get. Make the pinch when the plant is well established and 12 to 18 inches tall. Pinch just above a node where leaves are attached. I pinch the plant back to 6 to 8 inches. This pinch has the same results as the first cut described in Chapter 3; it is just being done earlier. Another time I have pinched with excellent results is when transplants could not be planted on time and begin to get overgrown in the soil blocks. As a last-ditch effort to save them, I pinched them while in the tray. Some of the sturdiest snapdragon plants I have ever planted had this treatment.

Which not to pinch: *Nonbranching* annuals, such as single-stemmed sunflowers and some varieties of cockscomb that are destined to give just one bloom should not be pinched. Do not pinch plants such as tomatoes and peppers because their central stem is the foundation of all the branching from which the fruit bears.

Pinching a 15-inch-tall cockscomb transplant to encourage branching. This will delay the first bloom but will bring more stems over time.

A lightweight floating row cover can be used with or without hoops. I use hoops if it'll be in place for more than a couple of weeks.

FLOATING ROW COVERS

I'm not sure I would want to garden without the security of a row cover. It can prevent many devastating garden mishaps. I use a lightweight cover that allows 85 percent light transmission. It can provide 4 degrees Fahrenheit of frost protection and is an excellent wind and pest deterrent. This lightweight cover can be laid directly on top of seedbeds for short-term use and used with hoops for a wider variety of long-term needs. The cover can be held in place with rocks or weight bags. When a heavier cover is needed, I double the lightweight cover. Here are the ways I use row covers. (Find suppliers in the Resources section for row covers, hoops, and weight bags.)

- **Planting seeds directly in the garden:** Lay the cover directly on the bed or use with hoops. Use caution in hot weather—laying a cover directly on a dry bed can toast seedlings. Covering will help retain moisture for better germination and protects seeds and emerging seedlings from birds, squirrels, and rabbits. Remove the cover once the seedlings have emerged and reach a size such that the bed can be mulched. Seedlings may bend under the cover but will quickly straighten up once it's removed.

- **Planting transplants in the garden:** Best practice is to hoop and cover. Covering will create a safe environment for the seedlings to become well established free from wind, hungry rabbits, deer, and digging squirrels and chipmunks. When planting in fall, winter, and early spring, transplants will especially appreciate the protection from grazing deer and cold wind that can quickly dehydrate foliage.

- **Protecting from pests:** Hooping and covering certain vegetable crops immediately after planting can delay or prevent a pest infestation. Preventing the pests access to the plants delays the pests' life cycle and can eliminate or greatly decrease damage. Remove the cover once blooming begins to allow for pollination. Some crops I have covered and protected with success include bush beans (Mexican bean beetle), squash (squash bug), strawberries (birds), potatoes (Colorado potato beetle), and eggplant (flea beetle).

- **Extending the season:** Using hoops and covers can allow an even earlier start in spring and extend the season further into fall with cool-season crops. Salad gardens and other cool-season low growers are excellent candidates for season extensions and even overwinter uses.

- **Using hoops or not**: Hoops lift row covers to create an air-space buffer under the cover in which plants can thrive. In winter, on bright sunny days, the heat is concentrated and keeps the soil a little warmer than the uncovered soil. This is also true in summer, so watch for hot days. I hoop and cover cool-season hardy annual flowers and vegetables that I plant in the fall all winter to keep the deer from grazing on them. It keeps the foliage in better shape, and the plants benefit from little growth spurts on sunny days.

Push one side of the hoop into the ground of the polyethylene tubing. Then do the opposite side. Hoops are sturdier when placed closer together.

Making and Installing Hoops

1. A roll of ½-inch polyethylene pipe can be found at building-supply stores. This pipe is flexible enough to form a hoop when it is pushed into the ground.

2. Cut the pipe into 60-inch lengths. Cut the ends at an angle to make pushing into soil easier.

3. Install a hoop every 5 to 10 feet of garden bed with the ends in the ground just inside the edge of the bed.

4. Push one end of the pipe into the ground, leaving the other end sticking into the air. From the other side of the bed, bend the pipe over and push the end into the ground opposite the other end. Pushing the ends into the ground 8 inches deep makes for the sturdiest hoops.

5. To install the cover, secure the ends first, pulling the cover taunt, using 3 weights on each end. Then place weights along the sides, one at each hoop. The goal is to prevent wind from getting under and lifting the cover.

6. Place the sandbag weights on the cover to keep it down in windy conditions, Sandbags can be filled with soil or use a heavy rock.

GARDEN IRRIGATION

Lay-on-the-ground irrigation is easy and inexpensive. If using any type of film mulch in the garden, the pictured drip tape is essential.

Permanent landscape and lawn irrigation has made using irrigation in a garden a little intimidating for many gardeners. The truth is that garden irrigation is very different and is easy and inexpensive to use. It typically lies on the soil surface and is easily installed with push and twist fittings. It is reusable from year to year. Irrigation saves water and puts it where the plants need it—at the roots, not up on the foliage.

The lay-on-the-surface drip lines or drip tape can be installed under film and mulch for the most efficient use. I use drip tape, a very cost-effective way to water a garden with straight rows. The drip lines are more flexible for use with plants that aren't in straight rows. (Find suppliers in the Resources section.)

ALTERNATIVE PEST CONTROLS

There are practices I follow to encourage unwanted garden guests to go elsewhere. When they don't, I eliminate them or what they want in my garden. There hasn't proven to be any flower or vegetable that has made me resort to taking measures that could potentially harm the ecosystem of my garden.

BLOCK PESTS FROM PLANTS

Crops that are known to attract pests, such as eggplants, potatoes, beans, and squash, can be covered immediately following planting. I prefer not to remove the covers at all until the plants outgrow them or blooming starts and pollination is needed. If you need to weed or tend the bed, replace the cover as soon as possible. This delays pests' exposure to the plants and their life cycle. Most often, only a few pests fly into the garden, but it's the eggs they lay, which later hatch, that cause the greatest damage. By delaying exposure to the plants, the outbreaks are delayed or prevented.

WILDLIFE DAMAGING FRUITS AND FLOWERS

If birds and other creatures are punching holes in your tomatoes and leaving the fruit, there is a good chance they are doing it for the moisture. When setting up and planting our tomato patch, we always include upside-down outdoor trashcan lids filled with water to train our birds and squirrels that this is the watering hole, not our tomatoes. It works wonderfully.

To prevent birds from eating from some fruit trees and shrubs, consider the timing of the harvest. We have fig trees on our farm that Steve's grandfather planted and have customer demand for every fig we can provide. I have found that harvesting them just before they begin to color up has prevented 100 percent of our bird damage. The figs quickly ripen indoors. Watch your fruit to see at what stage the birds are attracted and try harvesting before that stage. Bird netting is not an option I consider because of the high risk of snakes and birds getting entangled.

Make a Slug and Snail Trap

You will need a 12-ounce or larger plastic drinking cup, a 1- to 2-gallon plastic pot, scissors, a can or bottle of beer, a trowel, and a cinder block (optional).

1. Locate the trap in the garden near where the damage is occurring.

2. Dig a hole that will fit the cup and sink the cup into the ground so the lip of the cup is level with the garden-bed surface.

3. Fill the cup two-thirds full with the beer. Slugs and snails are attracted to the yeast in the beer and will fall in and drown.

4. Protect pets and other wildlife by covering the cup and the area around it with a plastic pot turned upside down. Cut several 2-inch notches in the lip of the pot as entrance holes for the slugs and snails. Turn the pot upside down and cover the cup of beer with the pot. I place a cinder block on the pot to keep my golden retriever from drinking the beer and eating the pickled slugs.

The slug and snail trap secured with a cinderblock to prevent nontargeted creatures (mainly a golden retriever) from falling in or drinking the beer.

In Japanese beetle season, the first job of the day is to head to the marigold patch where the beetles congregate for easy picking.

NO PESTICIDES—REALLY.

When I face an insect issue that seems to be getting out of control, I do have a couple of measures I am willing to take. Handpicking insects, blasting them with a stream of water, or offering a cup of beer to end their peskiness is sometimes in order.

Handpicking sizeable insects to squish is a method I perhaps enjoy a little too much. Stinkbugs, Japanese beetles, and giant grasshoppers are at the head of this list in my garden. Giant grasshoppers do so much damage so quickly that I have no mercy for them—they are easily

clipped in half with the clippers that are almost always in my hand while harvesting. The driving force in this practice is that each annoying insect I eliminate prevents many babies from being born the next season.

A trap crop for Japanese beetles is marigolds. Japanese beetles used to be a major cause of flower damage on my farm, but we rarely see them these days. During beetle season, I go straight to the giant marigold patch early in the morning with a little bucket of soapy water. The beetles dig deep into the blooms with their legs and tails sticking out. I place the bucket under the bloom and then tap the bloom. They react by dropping off the flower right into the soapy water. The soapy water prevents them from being able to crawl up the sides of the bucket and fly away. After I collect them all, I head to the bathroom for a flush. The most interesting thing begins to happen a few days after I start collecting them—the beetles hear my footsteps coming and begin to drop to the ground before I get there.

In an effort to reduce the number of aphids in a massive outbreak, a firm but gentle blast of water and at the same time rubbing the bugs off with your hand can do the job. I once washed an entire 100-foot row of butterfly weed that was being engulfed by aphids. This leveled the playing field and gave the beneficial insects an opportunity to catch up. The soft bodies of the aphids are easily damaged by water blasts, but be careful not to blast the good bugs.

Slugs and snails can wreak havoc on the garden. They can eat young plants to the ground in just one night. Even with all the leaf litter that I use in my gardens, we face little to no slug or snail damage. I attribute this to our garden being home to many slug predators, including lizards, snakes, foxes, common blackbirds, starlings, and large ground beetles. Fortunately, there is a home remedy to rid your garden of snails and slugs (see page 144).

DECORATIVE DEER FENCE

I once saw a decorative deer fence installed that kept the deer out and added attractive structure to the garden. It was made from cedar wood and was the required 7 feet tall with a barrier secured to the ground. Each fence panel was in a lattice pattern with approximately 6 inches between the slats, creating a very open appearance. It totally enclosed the yard and had a walk-through decorative gate. It was a beautiful solution to what was once an everyday problem.

OH DEER

I have successfully encouraged the small herd of deer that visit my garden to go over to my neighbors' yards, where they are met with more accessible plantings to eat. I have found the key is to have several discouragements and to rotate using them. And, most importantly, I must do something all the time.

- Deer sprays work when they are applied according to instructions and on the recommended timetable. Spraying should go on the calendar like a mortgage payment; if you don't miss a spraying date, the deer will stay out. Cornell University did a study that suggests rotating two or three different brands of sprays with different ingredients works best.

- Sprinkling blood meal, an organic fertilizer, on moist foliage does an excellent job deterring deer until it rains and washes it off. I used this method to run the deer out of our strawberry patch that was being eaten to the ground every night. I applied blood meal once a week and stopped them dead in their tracks.

- If you have strong deer pressure, a secure deer fence is the best prevention. There are many resources available on installing a deer fence, and I recommend doing your research before making any attempts to put up a fence. A nonsecure deer fence can lead to deer getting in but not being able to get out, or worse, getting caught in the fence and injured.

- If you have a dog, your pet can help. At dusk a couple of times a week, I walk the perimeter of our farm with our dog. She marks her territory and gives the deer reason to stay away. She has also been known to give a good chase and drive them over our fence, which makes them apprehensive.

I collect leaf litter from neighbors and dump it into my pathways; the leaves are not chopped first. Walking on them for a season breaks them down.

WEED PREVENTION IS KING

As soon as planting beds are prepared, they should be covered to block light from getting to the weed seeds that were brought to the surface when preparing the soil. Light pushes the seeds to sprout. When the bed is not covered, steps must be taken weekly to remove developing weeds to prevent the laborious task of pulling weeds later. Getting them when they are small is the secret to easy weed control.

Organic mulches do more than just make the garden look sharp and block weed seeds from sprouting. They also break down over time and feed the microorganisms in the soil. My first choice for mulch is always leaf litter. If using it on the bed top, chopping the leaves is beneficial. On paths and walkways, using whole leaves is fine. Other materials to use might be wood chips, pine straw, or other organic matter readily available in your region.

Laying several layers of unfolded and overlapping newspapers under organic mulch is a great way to further block the light from getting to weed seeds. The newspaper will break down over time.

Mulch Film Tips

- Lay any irrigation on the bed before laying film. Water only penetrates the film through the holes made for plants.

- The film should exceed the bed top and sides by 6 inches on all sides.

- Start at the head, holding the film down and snug. Work toward the other end of the bed.

- With the irrigation sticking out of the head end, cover the edges of the film with soil from the pathway to finish the head. Do the same at the opposite end with irrigation tucked under the film. All the edges of the film should be covered with soil, preventing wind from getting under the film and lifting it. The sun will tighten the film some over the first 48 hours.

- Use a dibber or screwdriver to easily poke holes through the film for planting transplants or seeds.

- Mulching or laying woven landscape cloth in the pathways surrounding the perimeter of the bed makes for lower maintenance.

1 *Cover the prepared raised bed with the film and use rocks or the like to hold the film down during installation.*

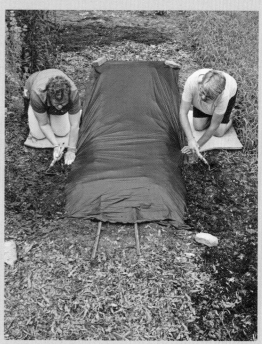

2 *Holding the film taut, use a hand hoe to scrape soil from the pathways over the edge of the film to seal the side edges.*

3 *Cover the end edge and then the head, making a hole for the irrigation to come through.*

4 *Finish the bed by mulching or laying landscape cloth to create weed-free surrounding pathways.*

Biodegradable mulch film is made from a corn byproduct, is biodegradable, and even adds a little nitrogen when it is turned into the soil. I find it easy to use. Over the years, it has contributed greatly to our weed-prevention program.

The film holds up for the life of an annual crop in the garden and can also be used under organic mulch in place of newspaper. It can be put down by hand or with a tractor and a mulch-laying implement. The film comes in black or white. The temperatures at planting time dictate which one to select to prevent cooking baby plants. When the high of the day is under 85 degrees Fahrenheit, I use black film. As the temperatures rise and the afternoon highs climb over 85 degrees Fahrenheit, I switch to white. The surface temperatures on the black are much warmer, which is helpful in spring and fall but deadly in summer for young plants. The white works well in summer to keep the surface temperatures down and preserve young transplants. Find suppliers in the Resources section.

HOW TO NEWSPAPER YOUR GARDEN

1. Unfold all the papers.

2. Fill a wheelbarrow halfway with water and submerge the papers. (Wetting the newspaper prevents the papers from blowing away while mulching.)

3. Take several layers of wet newspaper and place them on the garden bed, overlapping the layers by several inches.

4. Cover the newspaper with mulch.

Here, I am hoeing the fall garden where seeds were planted directly into bare soil. Using a hoe of the design that farmers use allows the gardener to stand upright. Pull the hoe's head through the soil like a razor. PHOTO BY GARDENER'S WORKSHOP FARM

CHAPTER 8

HOEING

When soil is left uncovered, weed seeds will begin to grow immediately. Running a garden hoe throughout the top 1 inch of soil like a razor will eliminate any developing weeds. By not going any deeper in the soil, hoeing will not bring up fresh weed seeds to sprout in the coming weeks. This task is quick and easy when weed seeds are just sprouting. Allowing weeds to get more mature will make it more laborious and will increase the time needed to eliminate them.

When seeds are planted directly in the garden into bare soil, weekly hoeing will keep the garden free from weeds. Run the hoe throughout the bed everywhere except where the seeds are planted. Once the planted seeds sprout and grow large enough to mulch, hoeing chores are done. The best part is that most of the weed seeds on the surface of the bed have been exhausted by the weekly hoeing, and there are virtually no weed seeds left to sprout. Find hoe suppliers in the Resources section.

NO-TILL VEGETATION REMOVAL

Woven landscape cloth is useful as a garden tarp. For a bed that won't be used for planting for 3 to 5 weeks, a garden tarp can smother worn-out plants, a cover crop, or weeds. This method does not disturb the soil or bring up fresh weed seeds to the surface and eliminates what is growing in the bed and turns it into residue from which the soil benefits.

Smother a Bed of Plants or Weeds

1. Remove any irrigation, netting, or other garden accessories.

2. Bend plants over or mow down to make them easier to cover.

3. Lay a black garden tarp over the bed with 12 inches of overhang on each side.

4. Place rocks or sandbag weights along the edges to hold the tarp down. Blocking the light and trapping moisture will aid the soil in consuming the plant residue.

5. This process can take 3 to 5 weeks, depending on temperatures. The warmer the temperature, the faster it happens.

6. Follow this treatment with an application of organic matter and organic dry fertilizer and the bed is ready to plant again.

To smother a bed of vegetation, the cover must exceed the sides of the bed and be secured to block light from getting to the soil.

COMPOST

Most gardeners face the same problem I do of not having the room or equipment required to make the amount of compost needed to maintain a working garden. A common subject of discussion when I speak to groups is that gardeners feel inadequate if they must purchase compost instead of making it themselves. Here's your pass—it is common practice for organic gardeners and farmers everywhere to purchase compost, so seek out reputable sources. It is available in bags at most nurseries and garden centers as well as in bulk from the places that sell bulk mulch.

While I purchase a good deal of compost each season, I still make compost even though it's not enough to sustain my garden. My compost bins are a lazy gardener's heap. Because I'm in the city and obligated to stay as tidy as possible and not give the neighbors anything to question, I build compost bins that look good and are easy to maintain. They do not get turned, so it takes a year or more to get compost. The compost is typically ready to use the following growing season. For a compost heap to cook effectively and kill weed seeds, it must be at least 3 feet wide and 3 feet tall.

I made this beautiful compost from the flower and vegetable debris the farm generates, but it's not nearly enough to sustain my garden. I must purchase compost also.

Large Garden Compost Heap

For yards and gardens that produce a large volume of vegetation waste, this large bin can be customized to the size needed.

This large compost bin made of straw bales will be overflowing by the end of the season. By next spring, it will be compost.

- The walls are built from straw bales staggered like bricks with air spaces between the bales.

- Usually two to three bales high does the job. One side stays open to wheelbarrow garden waste in.

- For our large volume of waste, the footprint is made of 10 bales on each layer, 4 across the back and 3 down each side.

- The open side faces into the farm, and the bale sides face the neighbors.

- We add all of our flower scraps and other garden waste but no perennial weeds or weeds with seed heads.

- The following season, we scoop the finished compost out and use it in our garden.

- If the bales are breaking down, cut the strings and use them to start the next season's heap. If the bales are still in good shape, start loading it again.

The smaller kitchen bin receives all the kitchen compostables, and I cover each deposit with straw from the bale I keep next to it.

Small Garden or Kitchen Compost Bin

I find that having a smaller bin closer to my back door for kitchen scraps is very convenient.

- To make a compost cage, use a piece of 2-inch-by-4-inch box wire fence, available at building-supply stores. Make a circle that is at least 36 inches in diameter. Secure the circle closed permanently with cable ties.

- Place the cage in an out-of-sight corner near the back door for convenience. To help keep the bin naturally moist, place it in dappled shade if possible.

- Place a bale of straw next to this bin to use as brown material to cover the kitchen waste as it is added.

- To keep varmints from visiting this heap, I make sure to chop kitchen waste into smaller pieces and don't add items that can take longer to break down, such as melon rinds and banana peels. They can be buried in the large heap or even in the garden.

- I empty my kitchen compost bucket into the bin and spread out the contents evenly, covering the kitchen waste with three times its volume in straw from the bale.

- The combination of kitchen waste chopped into smaller pieces and the straw quickly breaks down the compost. It is typically ready the following season.

- To get to the compost to use, lift the wire cage off (a two-person job) or simply cut the cable ties. I use any chunks or unfinished compost from the top and edges to start the next bin.

- Store the finished compost with a garden tarp over it to prevent erosion if you don't plan to use it right away.

LEAF MOLD

While compost adds organic matter and nutrients to the soil, leaf mold builds soil structure. Using both compost and leaf mold is preferred. What leaf mold does to build the soil's moisture-holding capacity and drainage qualities is matched by no other. If there is one secret to my gardening success besides persistence, it's leaf mold. Have as many leaf-mold cages as you can fit—they are easy to make. You just need space and time.

It couldn't be easier to make leaf mold. Dump leaves into wire cages and let them sit. If leaves are chopped first, they will process faster.

Making Leaf Mold

I tuck leaf-mold cages around the yard in unused corners. Typically, the lifespans of the leaf-mold cages overlap one another if you start one each winter.

- To make a leaf-mold cage, use a piece of 2-inch-by-4-inch box wire fence, available at building supply stores. Make a circle that is at least 48 inches in diameter and 48 inches tall. Secure the circle closed permanently with cable ties.

- Place the cage in dappled shade, if possible, to preserve moisture.

- Start filling the bin with leaves. We gather bags curbside and dump them in unchopped. (Chopped leaves will break down faster.)

- The cages are typically filled in fall, and the finished leaf mold is ready to use 18 months later.

- To access the material, lift the wire cage off the heap when it's ready (a two- or three-person job) or simply cut the cable ties.

- Add the finished leaf mold to garden soil. Unfinished leaf mold is excellent as mulch.

COVER CROPS

Cover crops are plants that are grown to benefit the soil and habitat and are not harvested to eat or use as a cut flower. Their worth is far greater and longer lasting than a mere harvest. Cover crops will attract and provide habitat for beneficial creatures, provide weed suppression, prevent soil erosion, and absorb excessive nutrients in the soil. In some cases, they add nitrogen to the soil, and they add lots of organic matter to it when they are turned under. They are commonly used on farms that have equipment to manage and turn the crop under, but they can also be used to benefit the home garden without equipment if the right crops are selected to grow and they are turned under in a timely manner. There are many other cover crops available beyond my suggestions. The two selected here are those that I have

A mix of warm-season cover crops: cowpeas and buckwheat. They provide habitat and blooms for creatures and will be turned into the soil to add organic matter.

used in small gardens, hand turned, and managed with success. Find suppliers in the Resources section.

Buckwheat for Spring and Summer Plantings

- Buckwheat is a fast-growing warm-season cover crop. It grows from seed to flowering in 30 to 45 days. The stem is hollow, so it is easily turned under and quickly breaks down.

Crimson clover is a cool-season cover crop that can be fall-planted to winter over in many areas. The blooms are a favorite of honeybees and chickens.

A visiting honeybee on buckwheat cover crop. This easy cover crop works well in small and large gardens, suppressing weeds with its quick growth.

- One pound of seeds covers 350 square feet. Scatter the seeds on the surface of the soil and then rake it in to cover the seed.

- Turn the crop into the soil while it is blooming and tender. If it's left in the garden beyond fresh blooms, it will make and scatter seed, which leads to reseeding buckwheat in the future.

- Buckwheat is a favorite of native bees, honeybees, and many other beneficial insects. Whenever I pull a flower or vegetable crop out of the garden and don't have a planting to go there immediately, I plant buckwheat seed. I plant it many times throughout the warm season to always have a patch of blooms available. Keeping buckwheat seed on hand is the secret to being able to plant it whenever a spot presents itself.

Clover for Late Summer and Fall Planting

- Crimson clover is a cool- to cold-season cover crop that will survive winters up to −10 degrees Fahrenheit. Planting 6 to 8 weeks before the first fall hard frost allows the seed to sprout and grow into a small plant to winter over, providing soil protection and habitat.

- Half a pound of seed covers 350 square feet. Scatter the seeds on the surface of the soil and then rake it in to cover the seed.

- Turn the crop into the soil while it is blooming and tender in spring. If left in the garden beyond fresh blooms, it will make and scatter seed, which leads to reseeding clover in the future.

- Crimson clover is an excellent spring source of food and habitat for native bees and honeybees. Chickens love clover as a special treat.

The Yard around the Garden

Our eastern-bluebird box is home to babies each season. They consume an enormous number of insects gathered in the garden.

Sunflower heads full of seed hang in a tree for the birds to feast on. The chickadees and finches especially appreciate this method.

Habitat is all about building community. Restoring a portion of the landscape beyond the vegetable and flower garden will go a long way in establishing a strong community of beneficial creatures. As the eco-system came to life in my garden, I began to wonder where these bugs and creatures went to spend the winter or where they went to nest and have babies. I started looking at the yard that surrounded the working gardens with new eyes and set out to make it more hospitable to all the good things I worked so hard to attract to the garden.

Not only would providing permanent habitat invite all these good guys to stay and hang out in my garden indefinitely, but their offspring would be more likely to stay also. I've learned that taking a few simple and inexpensive steps can go a long way in creating a place that these good guys would be happy to call home year round. Keep in mind that no pesticides—organic or otherwise—can be used in the surrounding yard without the potential of harming the beneficial insects and creatures for which you want to provide a safe place.

An untouched area in the wild island in my garden. Clover blooms provide for bees, and wild strawberries provide fruit for birds.

NATIVE-BEE HABITAT

Native bees need bare ground to nest. I have had to learn to leave certain areas free from mulch in the permanent planting areas surrounding the working gardens. In these same areas, I allow the naturally occurring plants (previously thought of as undesirables or weeds) to stay. While the goal in my working garden is to cover the soil with either leaf litter or biodegradable film, this bare-ground habitat is easily provided in the surrounding areas. To learn more about how to create a native-bee habitat, see Chapter 6.

A WILD SIDE

In the midst of my fairly tidy garden is a wild island. I've designated an area to remain untouched and let nature have her way with it, with the exception of removing invasive plants or saplings that would quickly grow into trees. It has become a hot spot of bird and insect activity and is rich in beautiful vegetation. It is easily kept neat and under control because the perimeter is mowed weekly, giving it a nice clean edge.

BIRD FEEDERS

I grew up in a home that had a bird ID book at the breakfast table. We watched many visitors come to a platform feeder my dad had made. My mother would fix jelly toast each morning for the Baltimore orioles that faithfully visited along with a horde of other birds. I learned that certain birds have food preferences, and when you provide for them, they became residents. With a little research into the type of feeder and food you offer, you can attract the birds you want and discourage those that may be bullies.

BIRD NESTING BOXES

Parents of baby birds are searching for insects 24/7 while the brood is growing up. The dimensions and location of a bird nesting box can make the difference of whether those baby birds grow up or not. Installing birdhouses with protection from predators such as snakes, raccoons, and other birds can make a big difference in raising a family successfully. Following the recommended dimensions of nesting boxes for specific species of birds offers protection and encouragement to them to use that box. The right box will be the most hospitable to the babies living in the box and will offer a way for them to be able to get out of the box when the time comes.

PROVIDE WATER

Everything that lives in your garden and yard needs water to survive—bees, birds, frogs, butterflies, hummingbirds, and others. If you don't provide it, they will leave looking for it. Drinking is very important, but bathing is also important, as this helps them maintain their feathers, wings, and skin. Plus, many of them simply appreciate a good swim. A pretty birdbath is nice, but the creatures don't really care what it looks like as long as they can reach the water to drink or bathe. Second-use items work well, including outdoor trashcan lids, larger saucers from pots, and really anything that will hold water. Different depths of water provide access for different creatures. Having a trashcan lid with a couple of inches of water with a big rock protruding from the water will provide for the bathers and swimmers as well as insects that can perch on the rock to drink. A daily ritual of dumping and refilling the baths will keep the water clean and free of developing mosquito larvae. This task is also a bit of a daily stress reliever for my husband, Steve, who puts down his stuff and refills the birdbaths as he walks to the house from his truck at the end of a work day.

The relaxing ritual of rinsing and refilling the birdbaths and anything else that holds water. Doing this daily will prevent mosquito larvae from developing.

Part of my native border. Clockwise from top left: southern wax myrtle, loblolly pine, viburnum arrowwood, common milkweed, Virginia mountain mint, and sweetbay magnolia.

CHAPTER 9

A place to perch, hide, and sun for creatures. Clockwise from top left: common milkweed, heliopsis 'Summer Night', black-eyed Susan, and Virginia mountain mint.

NATIVE PLANTS

Surrounding my working gardens with native trees, shrubs, and plants in the landscape has become a tremendous boost to the ecosystem of my garden. Why natives? Because they are the lifeline for everything else that is native—birds, mammals, insects, reptiles, and on and on up and down nature's food chain. Nature depends on native plants for a host of resources at many points and times in their life cycles. Why plant a non-native that may offer nothing more than shade and place to perch when a native can provide so much more?

I installed a 600-foot mixed native border on my farm a few years ago to provide a wind block for the gardens and to provide privacy from a coming housing development. Within months of planting, our population and activity of birds and other native creatures exploded.

ROCK PILES

Finally, I have a reason for a rock pile! I've always loved rocks but hadn't really found a purpose for them beyond laying them around the border of a flower bed. Rock piles provide crucial habitat for snakes, lizards, insects, and others if pockets are left open for them to crawl into. I find sunbathers stretched out in the heat of the afternoon, and this provides an excellent hunting ground for animals that eat said creatures.

Piles of brush in the native border provide cover from predators and offer a place to rest. The logs provide insects for birds to eat.

BRUSH PILES

A pile of logs, dead limbs, and brush creates cover for wildlife and insects. They will find refuge to rest, nest, and overwinter. It's the perfect place to hide from predators or take shelter in during severe weather. Start the pile with larger, longer limbs and logs on the bottom, then add smaller limbs and brush. Leaving pockets and entrance holes will encourage a busy pile. The logs and branches in the piles are also used by wood-dwelling native bees and insects that birds eat.

PERCHES

Most birds prefer the advantage of sitting up high and watching for insects and other prey. I have perches all over my garden. Some naturally occur, and others I installed. There are hundreds of garden stakes all over the garden. As they hold up the flower-support netting, they also provide easy perches for many looking for a meal. I also stuck some long bamboo sticks into our wire fences. Chickadees, eastern bluebirds, robins, and goldfinches use them regularly as perches. A row of 75-foot-tall gum trees flanking one side of the farm makes for very busy perches that also provide cover. On the open side of the farm, we installed a telephone-type pole that stands 35 feet tall to host a perch and our farm Christmas star. This perch is occupied during the day by eastern bluebirds and at night by great horned owls.

ALL THINGS WORKING TOGETHER

Including some of these habitat features, along with providing flowers in a pesticide-free zone, will have your vegetable and flower garden hopping with hungry creatures looking for pests. My garden has evolved over time. Don't expect these changes to happen overnight in a garden. They will come over time and be worth every moment of anticipation.

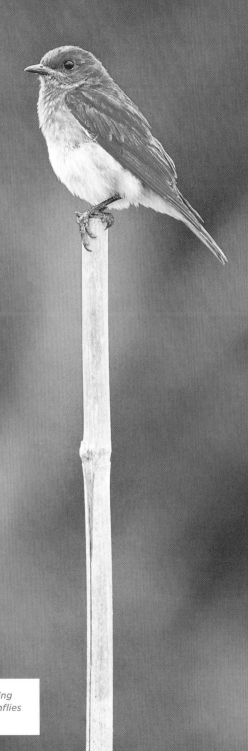

Tall bamboo stakes are pushed through the wire fencing around the farm every 30 feet or so. Birds and dragonflies love these high perches.

APPENDIX

About the Garden Plans

I offer two plans demonstrating succession planting for vegetable gardens with flowers: a six-bed large plot garden and a one-bed city salad garden. I have suggested plantings for each bed as it goes through spring, summer, fall, and winter. While seasons occur on different timetables for various regions, these examples are intended to give seasonal ideas that may be applied to your garden's timing.

- All beds are 3 feet wide and 12 feet long. Raised beds are recommended.

- The large-plot garden includes six beds. A 2-foot-wide pathway between beds and around the perimeter will make this garden approximately 16 by 32 feet.

- The city salad garden is a single bed.

- I recommend all flowers be spaced four rows across the bed with 6 to 9 inches between plants in the row. The one exception is sweet peas, which are planted in a single row with 6 to 12 inches between vines and located so they can be supported with a trellis.

- Vegetables are spaced four rows across the bed with 6 to 9 inches between plants in the row. The exceptions are tomatoes, cucumbers, and squash in a single row with 2 feet between plants in the row; peppers, which are planted in two rows with 12 inches between plants in the row; radishes, up to eight rows with 3 inches between plants in the row; and carrots, at six rows with 4 inches between plants in the row.

- Regarding bed #6: cool-season hardy annual planting time varies. Those in milder climates can fall-plant this bed. In northern regions, prepare the bed in fall and cover with mulch to be ready and waiting to plant in early spring. See chapter 4 for more on cool-season hardy annuals.

(A) Bush pea (mature, bearing)

SPRING

(B) Squash (mature, blooming)
(C) Pepper (mature, blooming)

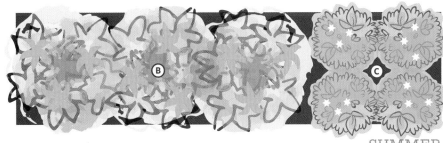

SUMMER

(D) Bush string bean (blooming)
(C) Pepper (bearing)

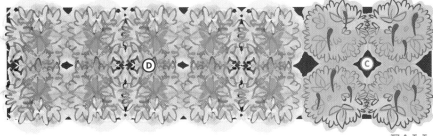

FALL

(E) Crimson clover (cover crop)

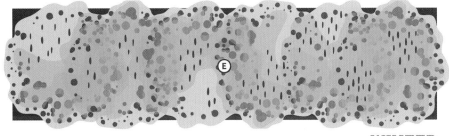

WINTER

BED 2

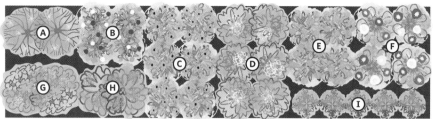

SPRING

- (A) Dill
- (B) Bachelor button
- (C) Rudbeckia
- (D) Snapdragon
- (E) Calendula
- (F) Poppy
- (G) Bupleurum
- (H) Monarda lambada
- (I) Sweet pea

SUMMER

- (J) Celosia plume
- (K) Cosmos
- (L) Celosia cockscomb
- (M) Strawflower
- (N) Zinnia
- (O) Sunflower
- (P) Basil
- (Q) Millet

FALL

- (J) Celosia plume
- (K) Cosmos
- (L) Celosia cockscomb
- (M) Strawflower
- (N) Zinnia
- (O) Sunflower
- (P) Basil
- (Q) Millet

WINTER

- (R) Clover (cover crop)

(A) Leaf lettuce
(B) Head lettuce
(C) Onion
(D) Swiss chard
(E) Beet
(F) Radish
(G) Carrot

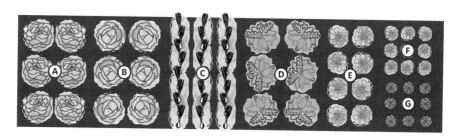

SPRING

(H) Buckwheat (cover crop)

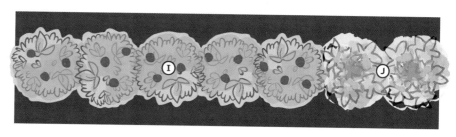

SUMMER

(I) Tomato (bearing)
(J) Squash (bearing)

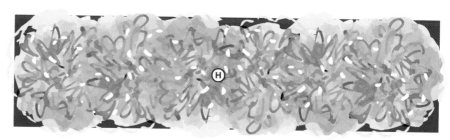

FALL

(K) Crimson clover (cover crop)

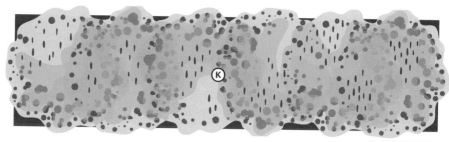

WINTER

BED 4

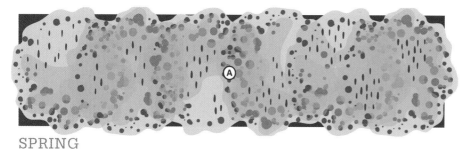

Ⓐ Crimson clover (cover crop)

SPRING

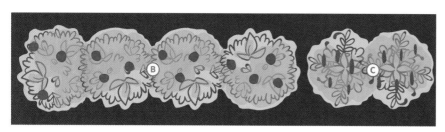

Ⓑ Tomato (blooming)
Ⓒ Cucumber (blooming)

SUMMER

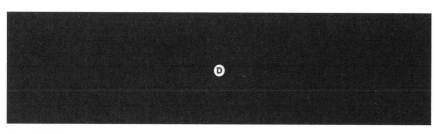

Ⓑ Tomato (bearing)
Ⓒ Cucumber (bearing)

FALL

Ⓓ *Bed prepared for early spring planting. Covered with mulch or biodegradable film through winter.*

WINTER

BED 5

- (A) Basil
- (B) Celosia cockscomb
- (C) Celosia plume
- (D) Zinnia
- (E) Zinnia (alternate color)
- (F) Sunflower
- (G) Ageratum
- (H) Cosmos
- (I) Marigold

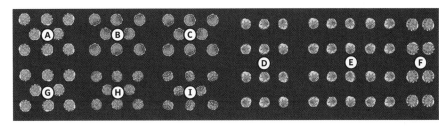

SPRING

- (A) Basil
- (B) Celosia cockscomb
- (C) Celosia plume
- (D) Zinnia
- (E) Zinnia (alternate color)
- (F) Sunflower
- (G) Ageratum
- (H) Cosmos
- (I) Marigold

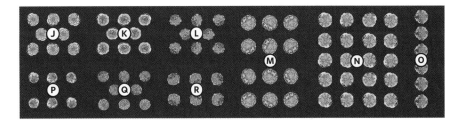

SUMMER

- (J) Poppy
- (K) Bachelor button
- (L) Monarda lambada
- (M) Snapdragon
- (N) Rudbeckia
- (O) Sweet pea
- (P) Calendula
- (Q) Dill
- (R) Bupleurum

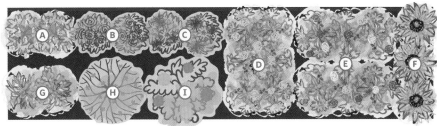

FALL

- (J) Poppy
- (K) Bachelor button
- (L) Monarda lambada
- (M) Snapdragon
- (N) Rudbeckia
- (O) Sweet pea
- (P) Calendula
- (Q) Dill
- (R) Bupleurum

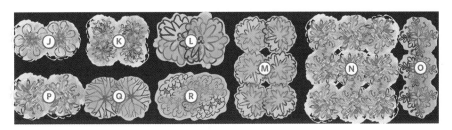

WINTER

BED 6

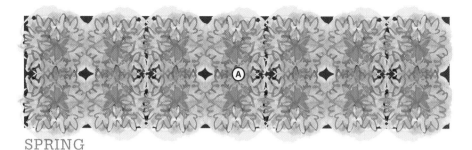

SPRING

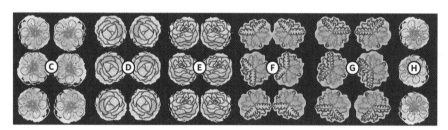

SUMMER

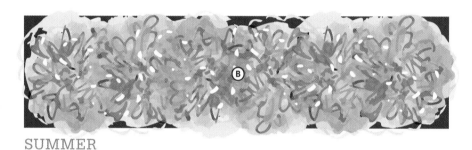

FALL

WINTER

(A) Bush bean

(B) Buckwheat (cover crop)

(C) Spinach
(D) Head lettuce
(E) Leaf lettuce
(F) Kale
(G) Swiss Chard
(H) Radish

(I) *Hoops and floating row cover installed to extend the harvest into cold weather.*

CITY GARDEN

(A) Bachelor button
(B) Calendula
(C) Leaf lettuce
(D) Spinach
(E) Head lettuce
(F) Onion
(G) Kale
(H) Zinnia
(I) Basil

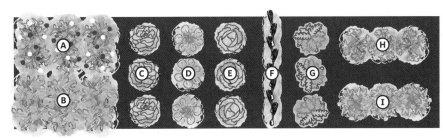

SPRING

(J) Sunflower
(K) Zinnia
(L) Tomato (blooming)
(H) Zinnia
(I) Basil

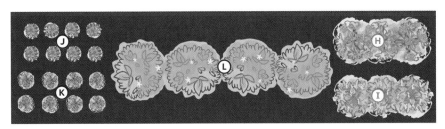

SUMMER

(J) Sunflower
(K) Zinnia
(L) Tomato (bearing)
(H) Zinnia
(I) Basil

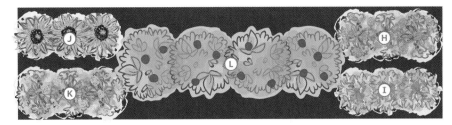

FALL

(M) Crimson clover (cover crop)
(N) *Bed prepared for early spring planting. Covered with mulch or biodegradable film through winter.*

WINTER

ACKNOWLEDGMENTS

My inspiration to write and teach comes from the gardeners, flower farmers, and cut-flower lovers whom I connect with at talks and online. Hearing about your love of growing flowers and the successes and challenges you've faced encourages me to continue—thank you.

The groundwork for *Vegetables Love Flowers* had been laid—I just hadn't recognized it—until that call came from Cool Spring Press's Mark Johanson. We began brainstorming, his insightful encouragement inspired this book, and the staff of Cool Springs Press brought it to life. Thank you all for believing in me and for this opportunity to share my message that flowers are so much more than a mere pretty face!

I've dedicated this book to my sister, Suzanne Mason Frye, because it could have never happened without her. Not only is she a big part of my life of farming, writing, and speaking—she is my best friend. I would have never pursued this project without her at my side taking care of me and my business. Thank you, sister.

The spectacular photos by award-winning nature photographer Bob Schamerhorn bring my garden to life on these pages. Even with my endless photo lists and 6:00 a.m. shoot dates—Bob was always kind, knew what I wanted, and delivered it. Thank you, Bob, for your patience, keen senses, and beautiful photos!

I am thankful to have connected with Shawna Coronado who introduced me to Cool Spring Press. Shawna has become a friend, and her willingness to nudge me is a gift I am grateful for; thank you, Shawna!

Without my good friend and personal editor, Susan Yoder Ackerman, I would have never dreamed of writing. Her continued encouragement, support, and editing give me the courage to do it.

There are so many with their hands in this book, and I am thankful to each one. My husband, Steve, is my biggest fan and the one who keeps our home running while I am writing. I couldn't do it without you. I love you, sugar—thank you.

RESOURCES

XERXES SOCIETY FOR INVERTEBRATE CONSERVATION
Learn more about pollinators and supporting them:
http://xerces.org/pollinator-conservation

JOHNNY'S SELECTED SEEDS
Cut-flower and vegetable seeds:
http://www.johnnyseeds.com

Bio360 biodegradable mulch:
http://www.johnnyseeds.com/tools-supplies/mulches-and-landscape-fabric

Cover crop seed:
http://www.johnnyseeds.com/farm-seed

Fresh cut-flower food:
http://www.johnnyseeds.com/tools-supplies/cut-flower-supplies

Row covers and hoops:
http://www.johnnyseeds.com/tools-supplies/row-covers-and-accessories

Soil blockers, heat mats, and grow lights:
http://www.johnnyseeds.com/tools-supplies/seed-starting-supplies

Standup garden hoe (6.5-inch Trapezoid):
http://www.johnnyseeds.com/tools-supplies/long-handled-tools

BERRY HILL DRIP IRRIGATION
Irrigation and Bio360 biodegradable mulch:
https://www.berryhilldrip.com

RAIN-FLO IRRIGATION
Sandbags, row covers, and irrigation:
https://www.rainfloirrigation.com

INDEX

MEET LISA MASON ZIEGLER

Lisa is an author, an accomplished speaker, and the owner of the Gardener's Workshop, a thriving small-market farm. She began her career selling cut flowers to local florists and Colonial Williamsburg, a business that soon grew to include florists, supermarkets, farmers' markets, a garden-share program, and a subscription service and later expanded into selling the tools, supplies, and seeds that she used in her own garden. At the same time, Lisa has steadily built a speaking career, leading presentations and workshops for garden clubs, master gardeners, commercial growers, and other groups centering on her simplified organic gardening methods.

Connect with Lisa on her blog, find her workshops, purchase her books, and more at www.TheGardenersWorkshop.com.

31901063546461